CAMBRIDGE LIBRARY COLLECTION

Books of enduring scholarly value

Technology

The focus of this series is engineering, broadly construed. It covers technological innovation from a range of periods and cultures, but centres on the technological achievements of the industrial era in the West, particularly in the nineteenth century, as understood by their contemporaries. Infrastructure is one major focus, covering the building of railways and canals, bridges and tunnels, land drainage, the laying of submarine cables, and the construction of docks and lighthouses. Other key topics include developments in industrial and manufacturing fields such as mining technology, the production of iron and steel, the use of steam power, and chemical processes such as photography and textile dyes.

A Treatise on Engineering Field-Work

In the 1840s, the civil engineer Peter Bruff (1812–1900) designed what was then the largest brick structure in Britain, the 1,000-foot-long Chappel Viaduct in Essex. He went on to become a railway entrepreneur and developer, and was responsible for the creation of the resort town Clacton-on-Sea, where he also designed many of the buildings. In this illustrated guide, first published in 1838 and here reissued in the revised and expanded two-volume second edition of 1840–2, he discusses the theory and practice of surveying (calculating the accurate position of points in the landscape) and levelling (calculating the accurate height of points). Volume 1 covers surveying; Bruff discusses different methods for calculating bearings and distances, and the equipment required. He explains the various errors to which each method is prone, how to avoid or minimise them, and gives example surveys of land boundaries, parishes and railway lines.

Cambridge University Press has long been a pioneer in the reissuing of out-of-print titles from its own backlist, producing digital reprints of books that are still sought after by scholars and students but could not be reprinted economically using traditional technology. The Cambridge Library Collection extends this activity to a wider range of books which are still of importance to researchers and professionals, either for the source material they contain, or as landmarks in the history of their academic discipline.

Drawing from the world-renowned collections in the Cambridge University Library and other partner libraries, and guided by the advice of experts in each subject area, Cambridge University Press is using state-of-the-art scanning machines in its own Printing House to capture the content of each book selected for inclusion. The files are processed to give a consistently clear, crisp image, and the books finished to the high quality standard for which the Press is recognised around the world. The latest print-on-demand technology ensures that the books will remain available indefinitely, and that orders for single or multiple copies can quickly be supplied.

The Cambridge Library Collection brings back to life books of enduring scholarly value (including out-of-copyright works originally issued by other publishers) across a wide range of disciplines in the humanities and social sciences and in science and technology.

A Treatise on Engineering Field-Work

Comprising the Practice of Surveying, Levelling, Laying Out Works, and Other Field Operations Connected with Engineering

VOLUME 1

PETER BRUFF

CAMBRIDGE
UNIVERSITY PRESS

University Printing House, Cambridge, CB2 8BS, United Kingdom

Cambridge University Press is part of the University of Cambridge.
It furthers the University's mission by disseminating knowledge in the pursuit of
education, learning and research at the highest international levels of excellence.

www.cambridge.org
Information on this title: www.cambridge.org/9781108071536

This edition first published 1840
This digitally printed version 2014

ISBN 978-1-108-07153-6 Paperback

A TREATISE

ON

ENGINEERING FIELD-WORK,

COMPRISING

THE PRACTICE OF

SURVEYING, LEVELLING, LAYING OUT WORKS,

AND OTHER

FIELD OPERATIONS CONNECTED WITH ENGINEERING.

With numerous Diagrams and Plates.

BY PETER BRUFF, C.E.,

ASSOCIATE INST. CIVIL ENGINEERS.

SECOND EDITION, CORRECTED AND ENLARGED.

LONDON:

SIMPKIN, MARSHALL, AND CO., STATIONERS' HALL COURT;

HEBERT, CHEAPSIDE; TAYLOR, WELLINGTON STREET, STRAND;
WEALE, ARCHITECTURAL LIBRARY, HIGH HOLBORN;
AND WILLIAMS, GREAT RUSSELL STREET, BLOOMSBURY.

1840.

LONDON :
CLARKE, PRINTERS, SILVER STREET, FALCON SQUARE.

TO

JOHN BRAITHWAITE, ESQ.,

CIVIL ENGINEER,

ENGINEER IN CHIEF OF THE EASTERN COUNTIES RAILWAY,
MEMBER OF THE INSTITUTION OF CIVIL ENGINEERS,
ETC. ETC.

DEAR SIR,

From my knowledge of your perfect acquaintance with Engineering Field Operations, and in testimony of the numerous acts of kindness which I have received at your hands while employed on some of the Public Works executed under your direction, I am induced to DEDICATE to you this *Second and Enlarged Edition* of a TREATISE ON ENGINEERING FIELD-WORK.

I remain,

Dear Sir,

Your obliged and very obedient Servant,

PETER BRUFF.

CONTENTS.

CHAPTER I.

CHAPTER II.

CHAPTER III.

CHAPTER IV.

CHAPTER V.

CHAPTER VI.

NOTICE.

Part 2 (which is far advanced, and will be issued as early as possible) will contain an elaborate Treatise on Levelling. Also a new Division which has been added to this edition on the subject of Laying out Works, &c.; the whole comprising a mass of *original* and *practical matter,* which it is presumed will be found useful to all engaged in the direction of Engineering Operations. The complete work will be very fully illustrated by diagrams and plates, and accompanied by a copious Index and Glossary.

ERRATA.—Page 35, last line but one from bottom, for "inner," read "outer." Page 47, for "plate 5," read "plate 4."

THE

THEORY AND PRACTICE

OF

SURVEYING.

CHAPTER I.

EXPLICATION OF THE TERM "SURVEYING."—SIMPLE CASE OF SURVEYING.—MEASURING A RIGHT LINE.—TAKING OFFSETS.

SURVEYING, in a general sense, denotes the art of measuring the angular and linear distances of objects so as to determine their several relative positions and draw a correct delineation of them, and to ascertain the superficial area or space included. It is a branch of applied mathematics, and supposes, in the operation, a good knowledge of arithmetic and the elements of geometry. As applied to the measurement of land, either for the purposes of computation, or for delineating the different natural or artificial objects which occur on its surface, it is performed by the measurement of several lines parallel to the horizon, passing in various directions, and which, being connected at their extremities, form some geometrical figure, either inscribing or circumscribing the object or space required. Thus,—suppose the following triangular space to represent a field, and the lines enclosing it, an open ditch: to survey such a piece of land, all that is requisite would be to measure the lengths of those ditches with some convenient unit of measurement, as a wooden rod of some ascertained length, as one, two, three yards, or more; or with a cord or chain of several of such units in length, instead of the wooden rod; and the operation of land-surveying in its simplest case will be thereby understood. Now, although the contents or superficial area of such a piece of land as that represented in the diagram could be easily computed on the ground, from the circumscribing measured lines, without drawing a plan or map of it, yet, for the purposes even of computation, as it will be seen, and of easy reference, it would be more convenient to lay down a plan of it on paper.

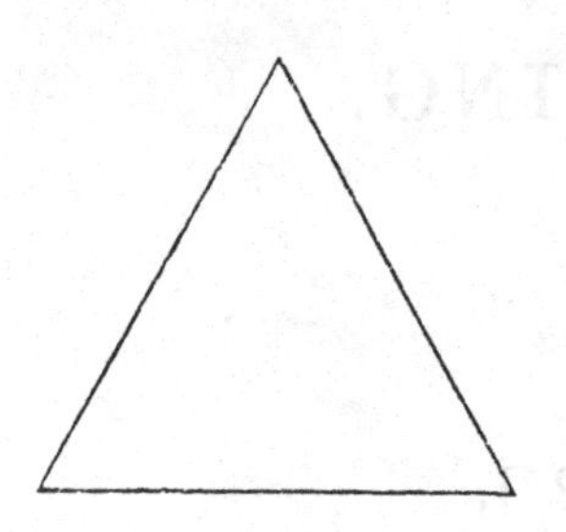

It will at once be evident that a plan of the ground cannot be drawn on paper to the natural or full size: a reduced or miniature copy only,—which should be one hundredth, one thousandth, or any convenient part of the full size, can be drawn, and the process is this:—Take a strip of paper or thin wood, divided into inches, which divisions might represent on paper the yards measured on the ground, if the unit with which the ground is measured be a yard; or each inch might represent two yards, or ten yards, or any number at pleasure. Now, supposing the measurement of the longest line in the diagram to be one hundred yards, that quantity might be represented on paper by one hundred inches, or fifty inches, or ten inches, or any value that may be assigned to the inches marked on the strip of paper or wood, which, thus used, is termed a scale. If the line on the paper is drawn one hundred inches, the scale would be one yard to an inch; if drawn ten inches, it would be ten yards to an inch, and so on. For convenience of measuring in the field, and for reducing the results of such measurement to the customary standard of acres, roods, and perches, which is universal in this country (except in land intended for building purposes, which it is usual to compute in superficial yards), a chain, twenty-two yards in length, is used, called "Gunter's chain," so named from the inventor, the Rev. Edmund Gunter.* This chain is formed of one hundred links, having a handle at either end for convenience of use. The links are joined together by three small rings, the whole thereby becoming very flexible. To measure with the chain, two persons are necessary, one to draw the chain, and another to follow, the last of whom, by a strange anomaly, is frequently called the leader, although we shall term him the follower. Accompanying the chain are ten iron arrows or pins, which, with the chain, are used in the following manner: —The point from which the measurement is to commence being determined, as well as the direction in which the line is to be measured, the leader takes one end of the chain in his right hand, passing his fingers through the handle of the chain and the eye of the arrow or pin, which he confines *within* the handle, but at its *extreme* part. Being thus prepared, he moves forward in the direction of the line to be measured, until he has drawn out the length of the chain; the follower, holding the handle at the other extremity, checks him as he draws it tight, and motions him right or left until his right hand, holding the chain, is exactly in the line, or straight to the

* By the use of this chain a considerable portion of the arithmetical calculations for finding the superficies in acres, roods, and perches, is performed in decimals. See description of the chain, and the computation of areas.

object to be measured to. When the follower has effected this, he holds the *outside* of the handle of his end of the chain to the starting point, and tells the leader to "put down," which order he obeys by fixing one of the iron pins in the ground, and immediately proceeds onward, drawing the chain with him, except he be desired to stop for the purpose of allowing the follower to take offsets, &c., the meaning of which will be presently explained. When another chain's length has been drawn on, the same operation is to be pursued, the leader holding the handle with the pin *within it*, and obeying the signals of the follower, moving right or left, as desired; the leader, in the same manner as at the commencement, holds the *outside* of his handle to the *outside* of the pin, and desires the leader, when in correct position, to put down another pin, at the same time picking up the one first put down, which should be hung on the *thumb* of his left hand. This operation is to be repeated until the whole distance to be measured is gone over, the leader putting down a pin at the end of each chain, which the follower, on arriving at, picks up, taking especial care *not* to pick it up until the leader has put down another. At the end of a line thus measured the number of chains will be denoted by the number of pins in the follower's hand, and the fractional part, if any, over and above the number of chains, by the number of links counted from the follower's end, the counting of which is facilitated by brass marks in the chain at every ten links, as will be hereafter described. If the measurement of a line should exceed the length of ten chains, it will be necessary, on the measuring out of the eleventh chain, for the follower (in whose hands the ten pins will then be) to go forward and put down a pin at the leader's end, or give him one for that purpose, after which, the remaining nine pins must be given up to the leader, the next one picked up by the follower being the eleventh: — In this way any distance can be measured, without difficulty or confusion, taking care to notice in the field-book each change of the pins, which, of course, represents ten chains. In describing the measurement of the above triangular piece of ground, it is presumed that the open ditches are straight from their extremities, and it should be distinctly understood that all lines must be measured straight from end to end. In the example just referred to, had the land been enclosed by a hedge or bank, instead of an open ditch, it would have been necessary to measure each side of the triangle a little *within* or *without* the bounding fence; in the one case the triangle formed by the measured lines would have been smaller, and in the other larger, than the enclosure; the quantity or distance of the measured lines from the boundary fence on each side, (for it is not necessary that the lines on each side the enclosure should be equidistant from the fences), would be determined at the extremity of each line, by a rod carried by the follower for such purpose, called an offset staff, the method of using which is easily explained.

It is evident that an enclosure bounded by crooked irregular fences cannot be defined by a single right line measured by the side of it. Its length can be determined by the measurement of a single right line, but the bends or crooks in the hedges cannot be so measured, although the omission of them would materially affect the measurement of the area or contents of the field. The bends in a crooked hedge or brook are determined by short measured lines, termed offsets, while a straight line is measured by the side of it to ascertain its length. The method of procedure is thus: — Let

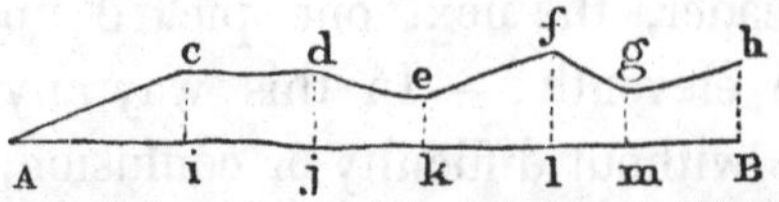

A c d e f g h be a brook or crooked hedge, and A B a straight line, measured by the side of it. From A measure towards B, stopping occasionally to observe if there are any bends in the fence. From A to c it appears to be straight; but, as it changes its direction at c, it is necessary to measure with the offset staff from the chain line, the length i c, at right angles to the chain line A B from the point i. The distance of i from A must be marked down in a book, as well as the offset i c; after which, the measurement of the line A B will be continued until opposite to another bend in the fence, as at d, where j d must be measured and entered in the book as before. In this manner as many offsets are measured as are necessary. To plot these offsets, or draw a plan of them, showing the crookedness of the fence, it is necessary to mark off, with the proper scale, the distances on the line A B, with the corresponding offsets thereto; a line drawn through the extremities of the offsets will represent the crooked fence A c d, &c.

CHAPTER II.

ON SURVEYING WITH THE CHAIN.—A TRIANGLE THE ONLY CORRECT FIGURE EMPLOYED IN SURVEYING.—DIRECTIONS FOR LAYING OUT LINES.—SURVEY OF A FIELD, DIAGONALLY.—DIRECTIONS FOR PLOTTING.—THE FIELD-BOOK OR REGISTER.—SURVEY OF A FIELD WITHOUT A DIAGONAL.—SURVEY OF A WOOD.—IRREGULAR SURVEYING.—SURVEY OF A ROAD.—MEASURING A LINE ON UNDULATING GROUND HORIZONTALLY.

THE observations contained in the last chapter having been carefully read, practical directions will now be given for surveying a single field with the chain alone.* In making a survey with the measuring chain only, we are confined to one geometrical figure—a triangle; for this reason, that, of all plane geometrical figures, it is the only one which, without altering the dimensions of its sides, cannot alter its form; † and, as in this case we only determine its form by the measurement of its sides, it follows that the correctness of the whole depends on the extreme accuracy with which the parts are measured, as well as in the judgment displayed in arranging or laying out the sides of the triangle on the ground. In the operation of surveying a single field, but little choice in this latter respect can be observed; but, where a choice can be had, the triangle should be as nearly equilateral as

* We say here "with the chain *alone*," because as the reader proceeds he will find other means accompanying the chain in effecting a survey. Reference must also be made to another part of this volume for an explanation of the various methods of computing the superficial contents of land.

† This follows from Euclid's proposition, "That upon the same base, and upon the same side of it, there cannot be two triangles having their two sides terminated at one extremity of the base equal to one another, and likewise their sides terminated at the other extremity."

possible (*i. e.*, equal sided, and, of course, equal angled), for, if either of the angles be very obtuse or acute, a very trivial error in the admeasurement of any one of the sides will materially alter the figure, and, consequently, the area enclosed within it. We consider it best to proceed gradually; we will commence, therefore, with a single field, as the same system is to be pursued throughout, whether it be a small enclosure, or a large estate, comprehending many enclosures, that is to be surveyed. The first operation to be done in the field is the arranging of the ground to be surveyed, into one or more triangles, according to its shape and circumstances. Thus,—suppose the field to be a four-sided figure; fix at the starting point, in one corner of the field on the longest diagonal or straight line that can be conveniently drawn from one angle to another, a conspicuous mark,* the top or visible part of which must be exactly over the point of commencement. Then look to the opposite corner of the field, and, if no natural mark can be seen sufficiently distinct and defined to be measured to, as a tree, the corner of a house, or other object on the diagonal line intended to be measured, erect a mark, as at the commencement; having done which, commence chaining from the first mark in the direction of the second, always observing to measure in a perfectly straight line. Leave marks on this diagonal or principal line, when arrived at D and E, opposite the other corners of the field, for the purpose of measuring tye or proof lines from those points to the vertex of each triangle, to verify the measurement of the sides. It must be noted in the field-book at what distance from the station or starting point these marks, or

* Generally a stick cut from the hedge, from four to five feet in length, as straight as can be had, with a piece of paper inserted in a slit in one end, from its appearance termed a "white."

"false stations," as they are termed, are put down. When arrived at the opposite corner of the field to that started from, put down another mark, if necessary, and from this station commence measuring a line by the side of one of the fences, without regard to the angle it makes with the preceding line beyond what we have previously mentioned, that it is to be neither very obtuse or acute. Offsets are to be taken to all the bends of the hedge as the measurement is proceeded with. At the termination of this line a mark is to be put down as before, from which a new line is to be measured to the first station. Then measure the tye line from the vertex of the triangle, or junction of the side lines, to one of the previous marks left on the diagonal, which will be a necessary and efficient check on the accurate measurement of the sides.

The same operation is to be repeated on the other side of the diagonal, and the survey of the field will be complete. The following diagram will at once show the manner of proceeding in surveying such a field.

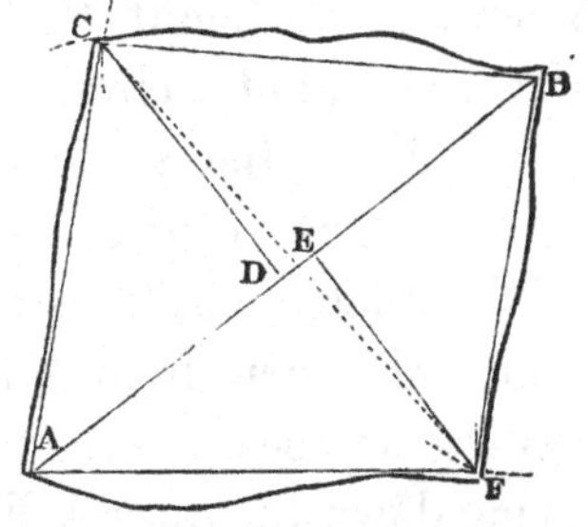

A B is the diagonal or base line; D E the false stations left when measuring the base line; B C one side of the triangle, commenced from the termination of the diagonal; and C A the remaining side, completing the triangle;—having arrived at the point from whence we started.—The measurement of the proof or tye line C D, however, remains to be done, to verify the measurement of the triangle A C B. The same method of procedure is also to be adopted on the other side of the line A B, measuring respectively the sides B F and F A, and the proof or tye line F E; or these separate measurements of the tye lines might be performed in one operation by measuring from C to F, as shown by the dotted line; taking care to ascertain the distance on the base line at which it crosses, as well as the distance from the same to the vertex of each triangle. This done, the measurement is completed. The next operation is, to lay down these lines (from which the field is to be plotted) on paper; to do which, select a scale to plot the field to, as one, two, three, or more chains to an inch, according to circumstances; but two and a half, or three chains to an inch, is the least scale from which quantities can be correctly computed.* The scale being determined on, draw the line A B in any position, and measure off the length or distance measured in the field with the scale, being careful to mark the position of the false stations D and E. Then take the length B C with a pair of compasses, and describe an arc of a circle from B as a centre. Also take the length A C in the same manner, and describe an arc of a circle from A as a centre. The intersection of these arcs will fix the relative positions of the lines A C and B C, to which point draw them from A and B. The same method of proceeding is to be observed on the other side of the diagonal, in laying down the lines A F and B F. When this is done, measure by the scale the distance from D to C; and in like manner from E to F. If these distances are the same on the plan as measured in the field, their agreement proves that the field-work was correct. If not, an error must have been committed either in laying off the lines on the plot, or measuring them in the field; in which case the

* The reader will bear in mind, we stated, in our first article, that land irregularly bounded was generally, and more readily, computed from a plan, than in any other way. The various methods of computing land are hereafter explained.

work must be gone over again, until it proves satisfactory. It is necessary, in even the most trifling survey, that a register or field-book be made, containing not only the extent of each line measured, the distance on each line where the offsets are taken, the lengths of the various offsets, &c., &c., but these particulars must be so disposed that no confusion or doubt may arise in the process of plotting. The accompanying register or field-book contains the necessary particulars, so simply disposed as that by no possibility can a mistake be made in plotting from it. The commencement of a line is denoted thus ⊙; and its termination by two double lines across the column of figures thus ═. F. S., by the side of a distance, means a false station, or, otherwise, a mark for reference, left at that distance. Principal stations are denoted thus ⊙. The sign ┐ denotes that the line was measured to the left, and ┌ to the right, of the line with which it is connected. The word *close* literally explains itself. The enclosure, of which the annexed column of figures is the field-book, is plotted to a scale of three chains to one inch, and is represented at page 9. It will be observed that the measurements are commenced at the bottom of the book, and

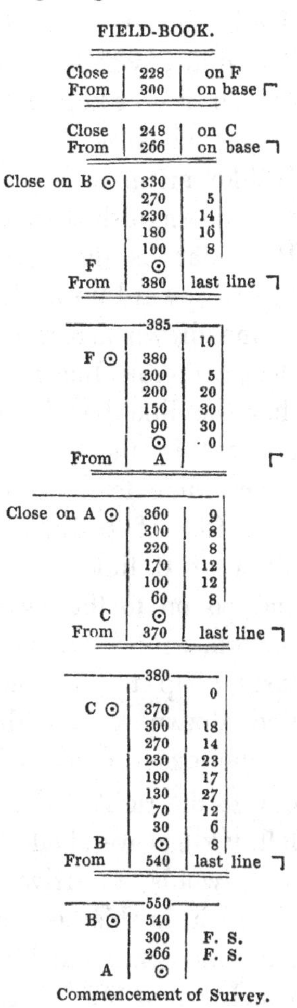

FIELD-BOOK.

Close	228	on F
From	300	on base ┌
Close	248	on C
From	266	on base ┐
Close on B ⊙	330	
	270	5
	230	14
	180	16
	100	8
F	⊙	
From	380	last line ┐
	385	
		10
F ⊙	380	
	300	5
	200	20
	150	30
	90	30
	⊙	0
From	A	┌
Close on A ⊙	360	9
	300	8
	220	8
	170	12
	100	12
	60	8
C	⊙	
From	370	last line ┐
	380	
		0
C ⊙	370	
	300	18
	270	14
	230	23
	190	17
	130	27
	70	12
	30	6
B	⊙	8
From	540	last line ┐
	550	
B ⊙	540	
	300	F. S.
	266	F. S.
A	⊙	

Commencement of Survey.

continued upwards, which is the universal practice in registers of land surveys, a plan which a few minutes' reflection will show to be quite natural. The chainage is always entered in links, a course which, without separately distinguishing the number of chains, prevents much of the confusion and uncertainty that would otherwise arise if chains or links had to be written opposite every entry of distances. Thus, in the third column from the bottom, instead of three chains being entered, we see 300 links, which is the same thing. Commencing at ⊙ and proceeding upwards, we come to 266, marked F. S., which is the distance to D on the diagonal A B. In like manner, F. S., at 300, signifies the point E. 540, distinguished by ⊙, is the termination of the line at B, so far as the figure is concerned, although it is continued onward up to the fence at 550, as understood by a line drawn across the column at that quantity; after that, a double line is drawn across the column, showing that the line has been carried no further. Immediately above this, the words "From 540, last line, ⅂," show a new line to the *left*. Commencing at ⊙ we have an offset of 8 links; at a distance of 30 links, an offset of 6 links; at 70, 12 links; at 130, 27 links; and so on to the termination of the line at 370,—although, as in the previous line, it is carried 10 links further up to the hedge, as shown by the single line drawn across the column at 380, which is also at the exact corner of the fence, as shown by an offset of 0 to the right. From 370 of last line we proceed to the left, taking several offsets up to 360, which is a close, or, in other words, an arrival at a previously determined point,—which, in this case, is A, from whence we started. The next line is measured to the right of A B, and from 380, of which, a new line is measured to the left, closing at 330 on the station at B. The line from 266 on base to the left is the proof line from D to the vertex of the triangle at C. The line from 300 on base to the right is the proof line from E to the vertex of the other triangle at F. A few examinations of the diagram, field-book, and description, will render this part quite plain to the reader; and, if he will take, moreover, the trouble to plot the enclosure a few times from the field-book to the same scale as the diagram in the first instance, but afterwards to a larger scale, he will have acquired as much knowledge in this particular as he will require, with the exception of actual practice in the field, which, in all cases, will be necessary to the practical understanding of the subject.

In surveying without a diagonal the outline of the field is plotted by determining the lengths of its sides and the angles; and the angles are determined by measuring the three sides of small triangles, each containing an angle of the field. Angles thus obtained are called chain angles. It is not at all an advisable method, although often practised to a considerable extent. There is a little time saved by it, and the chances of error are considerably multiplied. Circumstances, however, frequently occur in which it may be practised with advantage. In surveying a field by this plan, commence, as in the last example, at one corner, but, instead of measuring a diagonal to the opposite corner, measure a line by the hedge on the longest side of the enclosure, and when within two hundred links, or some other convenient number, of the distance intended to be measured out (which of course will be determined by the proximity and direction of the adjoining bounding fence), put down a mark. From the termination of the last line commence a new line along the side of the adjoining fence, but without regard to the angle it makes with the preceding line, except that it be as near sixty degrees, or the angle of an equilateral triangle, as circumstances will permit. When arrived at the same distance on the new line, as the former mark was put down from the termination of the preceding line, put down a similar mark; measure the distance across between these two marks, noting exactly the distance, even to half a link, or, otherwise, continue the line to the end of the fence, and measure the distance afterwards. The same method may be pursued at the other angles of the field, although it is not absolutely necessary to do so in a four-sided figure, except to *verify* the work, but the work should in all, even in the most simple cases, be verified, as, with the utmost care, errors may be committed, and, without a check, they will not be discovered until the operator finds himself in trouble.

The following diagram will illustrate this method of surveying a field, and the manner of plotting it:—

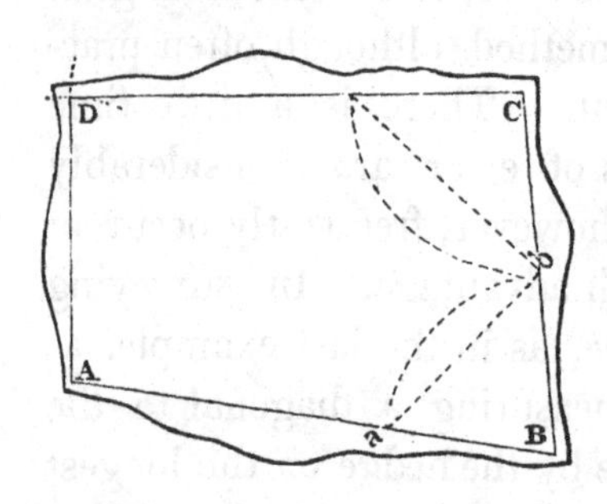

Commence at the angle A, and measure towards B, taking offsets where necessary; at two hundred links, or some other convenient number, before arriving at B, leave a mark a; when arrived at B, commence a new line, B C, and at the exact distance from B, on the new line, as a was left on the old line, put down the mark b, and carefully measure the distance b a; after which, continue the line onward from b to C, at which latter point it would be advisable to pursue the method just described, although, as we have before observed, it is not absolutely necessary in a quadrangular field or trapezium. The lines C D and D A, of course, are to be measured; and, if the point D be determined by a chain angle at C (by the same process as the angle at B), a check will be obtained on the correctness of the work when plotting, by having the measured distance D A. We think we have now sufficiently explained the method of surveying a field without a diagonal; what remains to be explained is the method to be pursued in plotting according to this method. First, draw the line A B in any position, and, with the scale to which it is intended to plot the enclosure with, mark off the distance from A to B; then take the distance from B to a, or B to b, which is the same, with a pair of compasses, from a much larger scale, and from B as a centre describe the arc a b; then with the compasses take the distance between a and b, as measured in the field. Put one foot in a, (observing also to take this distance from a large scale), and with the other make a fine puncture in b, or describe an arc cutting that point. Then with a fine pointed pencil draw in the line B C, passing through the point b, and observe to mark off the distance B C from the same scale as was A B. If the chain angle is also taken at C, the operation of plotting, or laying down the line C D, will be performed in the same manner as just described at B in laying down the line B C. But, if the lines C D and D A only are measured, the operation will be thus:—Take the distance C D with a pair of compasses, and from C as a centre describe an arc passing through D. Also take the distance A D, and describe another arc from the centre A, cutting the former one in the point D, and draw in the lines from C and A to the point of intersection, when the figure will be complete, and the offsets alone remain to be plotted, to determine the position of the boundary fences; but it will be perceived, by a slight examination of the diagram, that no check will thereby be obtained on the work, and any error committed will pass undetected. We should, therefore, to guard against error, take another chain angle at C (or any one of the three remaining angles) similar to that described at B; for, in measuring the diagonal a b, at the angle B, an error may easily be committed in reading off the chain, even when the greatest care is observed; which, if such should be the case, will remain undetected; for the two lines A D and C D, being plotted by intersection only, will appear to be right, however wrong they may be. But, if a chain angle be measured at C, for instance, there would be a check on the work, as the point D (in plotting) would thereby be determined; and, A being fixed by the first chain angle, it is evident the extremities of the line D A would coincide with those points (as previously fixed), only in the case where the work was correctly performed; for, if an error was committed, those points would be more or less distant than the length of D A.

It will often be found necessary to survey a wood or other enclosure without going into it, in which case the same method of procedure should be adopted as that last described, *viz.*, by chain angles, with the difference only of the angles being formed outside the figure instead of being within it, as in the former case. A right understanding of this method will be of great service to persons whose business may require them to make small surveys of property which the methods detailed in the preceding examples might not enable them to accomplish; where, for instance, instead of a wood, as represented in the diagram, houses and gardens, with high walls intervening, are supposed to be within the boundary fence; it would not only be inconvenient in such a case but almost impossible to measure lines within the boundary.* The accompanying diagram represents a

* This and the preceding method of forming a trapezium with the chain alone, is the only way in which a town survey (comprehending many streets)

wood, through or within which lines cannot conveniently be measured: means must therefore be adopted of determining its extent and form of boundary by measuring lines around it, *without* the boundary.

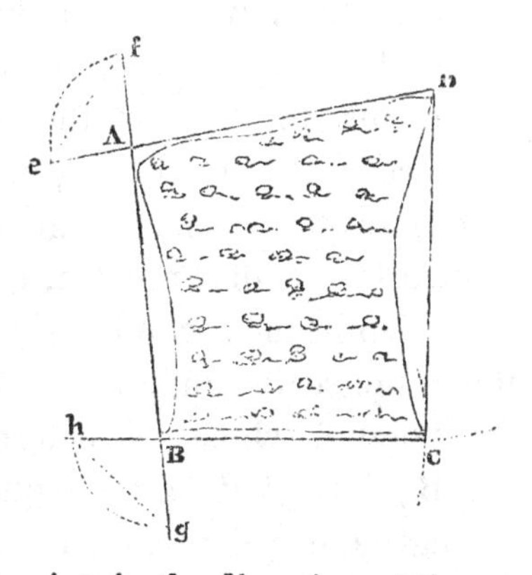

First, determine on the position of the lines A B and A D, so that their measurement may be accomplished clear of all obstacles. When this is arranged (either by fixing up marks at B and D, or taking natural objects, if such exist, in the direction of those lines), retire back to f, on the line B A—say one hundred, or some other convenient number of links,—and mark the spot. Set up a mark at A that shall be visible from D, and measure the line A B, taking offsets to the fence as the measurement proceeds. When arrived at B, or at such a point that a line, B C, can be measured conveniently up by the side of the adjoining fence, put down a mark, B, and continue the line onward to g,—say one hundred links, which is the same distance beyond B as f beyond A. From B retire back to h, on the line C B, the distance B h being the same as the distance B g, and measure the distance across from h to g. Measure the lines B C and C D, and when arrived at D, the mark previously set up at A will be visible; the line D A can then be measured without difficulty; in doing which, carefully note in the field-book the junction or close of the lines at A, but

can be made without the aid of angular instruments; but, in making a town survey by this method, it should be carefully borne in mind always to get the greatest possible diagonal in taking angles with the chain. We need scarcely add that angular instruments must be used in all cases where great accuracy is required.

continue the measurement on to e, the same distance from A as the point f was previously put down; measure e f, and the survey will be completed.

To plot such a survey as that just described, draw a line at pleasure to represent the line e A D, and mark off the distance A D with the required scale, then with a pair of compasses take, from a larger scale than the survey is to be plotted to, the distance A f, or A e; and, with one foot in A, sweep in the arc, e f. Also take the distance e f from the large scale, and, with one foot in e, bisect the arc at the point f, which will determine the position of the line f A B; but great care is requisite in drawing it exactly through the points f and A, as the slightest deviation from these points will materially alter its position at B. When this line is drawn in, lay off the distance A B with the proper scale, and extend the line onward to g. Take the distance B g, with a pair of compasses, from a large scale, and from B describe the arc g h. Take the distance g h from the same large scale, and from g as a centre describe an arc cutting h; draw in the line h C, passing through B, and lay off the distance B C with the proper scale, when the point C will be determined. It only remains to draw a line from C to D, to complete the figure, which distance on the plan, as determined by the scale, must correspond, if correct, with the distance measured in the field. If accuracy in the measurement of the chain angle at A, and of the circumscribing lines could be insured, there would be no occasion for the chain angle at B; but, as all operations, both of measurement and calculation, are doubtful if not checked or proved, it should be an invariable rule in all operations connected with surveying to obtain a check on the work.

The surveying a road with the chain only is performed thus:—Suppose A B C D E to be a piece of road requiring to be surveyed, and no other means at hand than the chain to operate with: in such a case, commence at A; and, if no mark is visible on the line A B, set up one. Observe to get the longest possible sight along the line of road, and take offsets to the bends in the fences on each side, as the measurement proceeds; likewise to all houses, buildings, and other objects of importance, within offset distance; and to all fences running up from the road. Before arriving at B it will be seen if the angle A B C

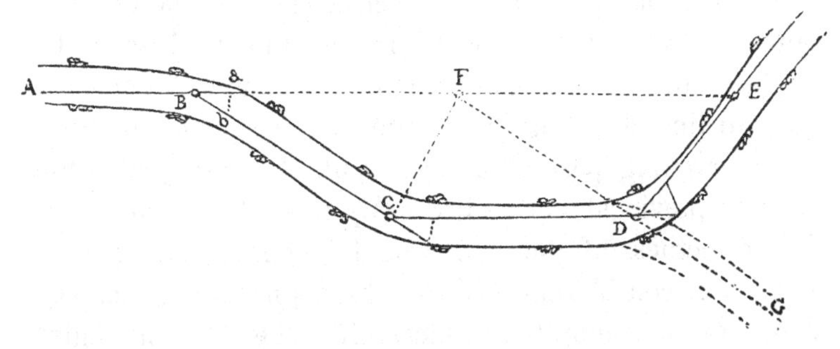

is acute; if so, set down a mark at fifty or one hundred links from it: but, if the angle is obtuse (as in the example), continue the line beyond B to a. Commence a new line from B, and measure to C, and at b, the exact distance from B as the mark a was previously put down, leave a mark, b, and carefully measure the distance across from b to a (see note, page 16); then continue the line onward to C, on arriving at which, pursue the same method for determining the position of the line C D as just described for the line B C, which method is also to be pursued as often as new lines are required. In plotting the road, the same method is to be pursued as directed for plotting a field surveyed without a diagonal. This method of surveying a road has not much pretensions to accuracy, as the small angles that can be formed with the chain, in a road of ordinary width, are not accurate enough, even with the greatest care, to *insure* the correct position of the various lines, by which the features of the survey have to be delineated.

Such a piece of road as that represented in the diagram might, however, be surveyed pretty accurately with the chain in the following manner: — Continue the line A B to E, where it will intersect the road; determine the direction of the forward line from E, in the usual manner by a chain angle, as also the direction of the backward line to D; but generally in such cases the station at E can be so placed that the backward line to D may be produced through E, and continued on so as to become the forward line, avoiding by such means one chain angle:— the intervening portion of road would be surveyed in the manner previously described. By this plan a check on the correctness of the work would be obtained on closing at E; for, considering A E as a base line, any inaccuracy that may be committed in measuring the sides or angles of the trapezium B C D E would cause the extremity of the line D E to fall short or beyond the proper point. The measurement of the chain angles might be altogether omitted if a station was formed at F, and lines measured to C and D, which would resolve the whole into three triangles—mutually checking each other. If the road had turned to G from D, instead of to E, the dotted continuation of the line F D shows the position of the new road line. In all cases of road-surveying with the chain, if the several lines be extended for many chains beyond the limits of the road, and their extremities carefully connected (see diagram, page 17), an accurate triangulation might be effected;—but in all cases of this kind much more correct results will ensue from obtaining the direction of the lines with an angular instrument, as hereafter explained.

A line measured on undulating ground requires reducing

to the horizon before it can be accurately laid down on a plane surface, such as is a map or plan. It will be evident, to the thinking reader, that, in measuring the distance between two objects, between which a hill intervenes, if the measurement is carried up on one side and down on the other, the distance so determined will be greater than if no such hill intervened, and the measurement between the two objects was made on horizontal or level ground. The excess of distance will be the difference between the hypothenuse and base of a right-angled triangle. At present we shall not stop to explain what the excess of distance is (which of course varies with the inclination), but proceed to explain in what manner it can be practically remedied when the ground is but moderately steep; *viz.*, by taking short lengths of the chain, and holding it horizontal. If the hill is not very abrupt (such that a man can walk up without difficulty), half the chain may be used at one time in continuing the measurement forward. Supposing this to be the case, the leader, in ascending a hill, would hold the handle of the chain quite down to the ground, while the follower would take fifty links, or half the length of the chain, and hold it up over the last pin until the chain was as near level as he could guess. The leader would then put down a pin, when another half chain would be measured out in the same manner, and the preceding pin returned to the leader; as, although two pins would be put down, only one chain forward would be measured. The returning of the pins must be particularly attended to at each second repetition of the half chain, or mistakes will inevitably occur; or, if it be preferred, the chainage up to the foot of the hill may be entered in the field-book, and all the pins given to the leader. The pins put down at each half chain may then be kept until the termination of the slope, when of course half the number of pins taken up by the follower will be the number of chains to add to the distance entered in the field-book. We have yet only spoken of the ascending of a hill; the only difference to be observed in descending is to reverse the operation, the follower holding his end of the chain to the ground, and the leader elevating his end until he thinks it horizontal, and then dropping his pin. This method of correcting the chaining over undulating ground cannot but be attended with much uncertainty, entirely depending, as it does, on the judgment of the operator in holding the chain level. It is also evident that, if the difference of level of the ground in half the chain's length exceeds the height to which the operator can hold the chain, the distance must be reduced. We have seen as short lengths taken as ten links; but no dependence can be placed on the results, where the ground is of such a steep character as to require so short a length of the chain to be taken.

Many surveyors, in adopting this method of measuring over undulating ground, have a plumb line, which the man elevating the chain holds in his hand, by which means he can ascertain pretty correctly, when ascending, if the end of the chain is over the mark, and, in descending, where to put in a pin. Such a method may be practised pretty successfully on slightly undulating ground; but, if possible, recourse should be had to angular instruments, by which means the angle formed by the slope with the horizon can be measured, and the necessary deduction made from the distance measured parallel with the surface of the ground; or, otherwise, by ascertaining the difference of level of the ground, and having the *surface* measurement, a simple calculation will give the necessary deduction. The latter part of this article, however, properly belongs to another and more advanced part of our work, where the reader will find full information on the subject.

CHAPTER III.

GENERAL DIRECTIONS FOR CONDUCTING A SURVEY.—LAYING OUT A BASE LINE.—FIXING STATIONS.—ERRORS COMMON TO CHAIN SURVEYING.—CHAIN ERROR.—STATIONS AND STATION MARKS.—OFFSETS, LENGTH, MEASUREMENT, AND ECONOMY OF.—BOUNDARY FENCES.—ERRONEOUS PRACTICES IN MEASURING RIGHT LINES.—AVOIDING OBSTACLES IN SURVEYING.—LAYING OUT FENCE LINES.—REMARKS ON FIELD-BOOKS.

In the preceding chapter having given the reader what we consider sufficient elementary instruction in the principles of surveying, we proceed to give some general directions for the conducting of an ordinary survey.

In the surveying of tracts of land, however extensive, the same principle should be followed out as observed in the survey of a single enclosure (such as given at page 9), with this modification,—that the tye or proof lines, instead of being measured exclusively for such purposes, are generally so arranged as to take up important features on the ground, such as it would be necessary to represent on the plan.

Surveying, as we have previously observed, may be performed in various ways; with the chain only, as just described, or by means of angular instruments, combined with the chain. It should be here particularly observed that the base or principal line of a survey, from which all the other lines diverge, should, if practicable, be carried through the greatest extent of property to be surveyed, so as to intersect, or pass near to, the principal and most intricate parts of the work. It is also well to carry the base line near midway through the property, leaving nearly the same quantity of work to be done on the one side the base as on the other. Then, the tye lines, in filling in, should cross the base and tye into the opposite side of the figure into which the lines are resolved; which, in plotting, will be a satisfactory test of the accuracy of the work, and thereby insure the correct positions of the several objects to be delineated in the survey. We may here also remark that too much care cannot be bestowed on the laying out and measurement of a base line, for on the correctness of such fundamental line every part of the survey will depend, whether trigonometrical, or plain surveying with angular instruments and the chain, or the chain alone. Any inaccuracy in the base line will inevitably distort the whole survey, and alter the position of every object that may be delineated.

Stations of all kinds should, if possible, be made at *even* chainage, *i. e.*, either at the extremity of the chain, or at one of the decimal divisions denoted by a brass mark (see article on the chain); or, if this cannot always be done, take the point midway between, which is five. The reason for this is, that, plotting scales being divided into tenths and twentieths of a chain, the distances at which the stations are put down can be more correctly marked off, in plotting, from one of the divisions on the scale coinciding with the intended station.

Chain surveying (in which the entire accuracy depends on the intersection of measured distances), is much more limited in its capability than when combined with an angular instrument, and certainly not so correct. In surveying with a chain, we are, in every primary case, limited to one figure—a triangle; and the correctness of the survey will entirely depend on the extreme accuracy with which the sides of such figure are respectively measured. Precipitous or enclosed ground, with strong fences, stone walls, or earthen banks, will render it almost impossible to measure distances with the requisite degree of accuracy; and an error in the measurement of the side of a triangle, if only a few links, will not be confined to the side on which the error is made, but will extend itself to the whole figure, and alter the position of every object enclosed within it. Not so, if the direction of the sides be determined with an angular instrument; for, in such case, if the angles are correctly taken, and afterwards protracted on the plan, the sides will be placed in their true position, and the admeasurement with the chain of the sides will, on plotting the work, determine the correctness or incorrectness of the angles. By this method the angles and measurements become mutually checks on each other.*

The correctness of what we have just stated will probably be made more clear by means of the accompanying diagram, in which A B C represents a great triangle, the sides being determined only by measurement, and their position consequently dependent on intersection from the extremes of any one of the sides, which may be taken as base, and assumed to be correctly measured, and which in the diagram is represented by A C. Now supposing the lines A B and B C to be

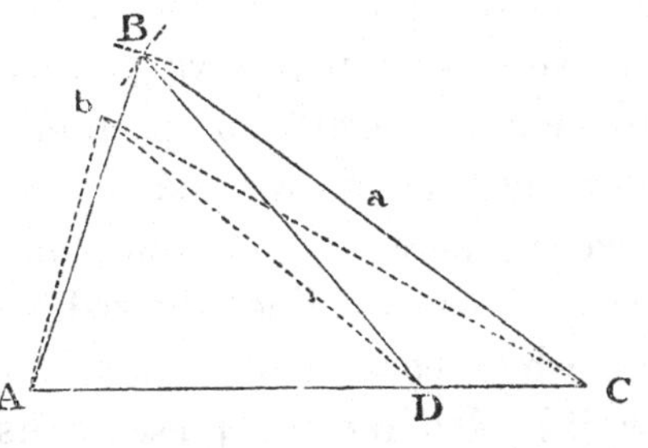

* The correctness of the angles may be proved without trouble, by simply adding together the three internal angles observed at the vertices of any *plane* triangle, which, if correct, will in every case amount to 180°, into which a semicircle is divided: if the figure should be a trapezium, the four observed internal angles will together amount to 360°, into which the circle is divided. For a further elucidation of this matter, see method of surveying with the sextant, theodolite, &c.

correctly measured, their intersection will determine the point B; but if an error be made on the line A B equal to the distance b B, it is evident the vertex B will be brought down to b, and yet the whole appear to be correct. But the tye or proof line, B D, would be a supposed check on the correctness of position of the point B; it will, however, be found that the distance D b is precisely the same as the distance D B; and, notwithstanding the tye line apparently proves the position of b to be right, we perceive that it is incorrect. Certainly, it may be said that the position of the proof or tye line is too oblique to the base, which we fully admit, but it sometimes happens, from the features of the country, that no choice is left to the surveyor, and he is obliged so to lay out his lines. Again, it may be said that a line measured from about the point a to D would point out the error; but a surveyor does not always perceive the most advantageous direction in regard to position in which to run his lines until he draws the diagram on paper. We will, however, here suppose that, on plotting the above diagram, the proof line showed that an error had been somewhere committed; but where is the surveyor to look for it? No remedy presents itself but to recommence the work, and measure over all the lines again from the commencement, until the error is detected. But, had an angular instrument been used in determining the position of the several lines, it is evident that no such mistake could have arisen, or, if so, that the mistake would have been immediately detected, and its rectification consequently become an easy matter. Supposing the angles C A B, A B C, and B C A, be taken, it is evident that the angles C A b, and A b C are, on the plan, greater than could be those angles observed in the field, and the angle A C b less, whereby the error would be immediately detected; for suppose, as before, the three angles to be correctly observed and plotted, but a mistake to have occurred in measuring the line A B, equal to the quantity b B; the direction of the line C B would be correctly determined by the angle, and the point B by the measurement; the direction of the line A B would also be correctly determined by the angle, but the point B, by the erroneous measurement, would appear to be at a point near to b on the line A B, immediately showing that the line A B was too short, and that a mistake had occurred in measuring it, equal to the quantity b B.

A further source of danger in trusting implicitly to the measurement with the chain of the sides and proof line of a triangle may be exemplified in the following diagram:—Suppose A B C to represent a great triangle, and B a to be the proof or tye line; now, after laying down the triangle (of course by intersection, for we have supposed no angles to be taken), the measurement of the proof line a B by the scale on the plan being found to be the same as the measurement in the field, the operator would generally be satisfied of the correctness of the whole; but there would be no certainty of it; for the tye line being right-angled, or nearly so, to the base A C, it would plot almost as well, or perhaps better, when drawn to b or c, than to the proper point, which might arise from many circumstances; such as an error being committed on the lines A B, or B C, or from the ruggedness of the ground over which B a was measured. We therefore perceive that we have no proof of the correctness of the triangle; for,

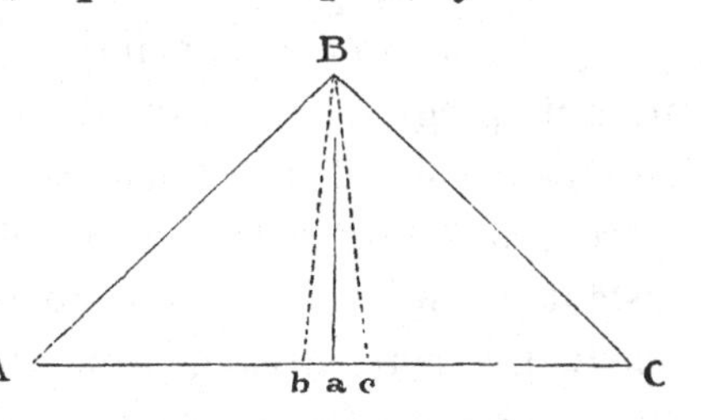

an error committed in the measurement of any of the lines, or a wrong entry in the field-book, may place the station a either at b or c; and this liability to error, but apparent correctness, will always attend measurements when meeting at or near to a right angle, without the direction of the lines be determined by an angular instrument.

The various difficulties and obstructions which often occur to the preventing of correct measurement of right lines — such as woods, buildings, water, and the like, will be found treated of in another part of this work.

An important point to be attended to by every surveyor previous to commencing a survey, and occasionally during its progress, is, frequently to test the accuracy of his chain, as, from the continuous strain, chains, if made ever so strong or heavy, soon become elongated, which, without constant attention, will lead to very serious errors. To remedy the evils arising from the tension of the chain being over great (which causes its elongation), it has been proposed to fix a spring balance at the follower's end, to regulate and make uniform at all times the degree of tension. But we conceive much trouble would be thereby occasioned without adequate correct results being obtained.* Occasional measurement of the chain, and some slight care, are all that is required to insure the surveyor from errors

* In measuring a base line where more than ordinary accuracy is required, a dynamometer attached to one end of the chain would be desirable, and so constructed as to show the degree of tension to which it was subject at each repetition of the chain, and which, of course, should be uniform. We understand that the standard chains furnished by the tithe commissioners have such an appendage. In the measurement of the base line, for the trigonometrical survey of England, by Col. Mudge, on Hounslow Heath, a steel chain of 100 feet in length was used, constructed similar to a watch-chain, which was supported throughout its length in wood troughs, placed horizontally, and kept at a constant and uniform degree of tension by a weight of 14 lbs. suspended from one end. A weight of 28 lbs. was subsequently used for that purpose, in measuring the bases of verification.

arising from such source. It is customary for the principal tithe and poor law surveyors to be furnished with standard chains to test the correctness of their ordinary chains; and they are directed to set off the length of the standard in some convenient level spot, and mark the extremities for the purpose of daily comparison. Where the surveyor has a standard chain, of course little trouble will be experienced by a comparison of lengths; and, even where such facilities are not at hand, the correct length of a chain can be easily laid down on the ground from a rod of five or ten feet in length, carefully divided into feet from a plotting scale, or carpenter's rule. At each extremity of a chain's length so determined, a square stake should be driven, the outside of the handles of a correct chain, when drawn tightly, barely reaching to the inner sides of the stakes: this is a much better method than that in current use, of making the chain's length from centre to centre of the stakes. When a chain is discovered to be too long, the amount of error, when considerable, should be distributed between the ten divisions of the chain, and the proper quantity taken from each; which should be done by the removal, from each division, of one or more of the small rings which unite the links. If the amount of error is small, it is customary to correct it at the extremities of the chain, bearing in mind that it must be so corrected equally at each end; if so rectified, the centre will occupy its right position, and the remaining divisions of the chain will be so triflingly in error as to be beneath notice. The error, however slight, which may exist in a chain's length, will be very dangerous if the main lines of a survey be measured with it. We cannot too earnestly impress on the surveyor's mind the necessity of measuring his chain not only at the commencement of a survey, but frequently to compare the same with the standard length during its progress, as we have observed that it is generally, we might almost say universally, neglected; most practitioners being satisfied with two or three comparisons in a year, although the chain be in constant use. Further, to show the necessity of daily comparison of the chain with the standard, especially while the principal lines are being measured, it need only be remarked that, if, between the days of reference, the chain should be found to have stretched any small quantity, it might be safely allowed for on the day's work by adding one-half the error in one chain's length to the total number of chains measured; but, as the chain, after being for some time subject to the tension which caused it first to stretch, will cease to stretch further with such strain, it will be evident that, if after several days' work the same correction should be applied, it will be uncertain and erroneous.

Stations are invariably put down at the crossing of every fence—generally on the ditch side, if equally favourable with the other side, for measuring a line, to offset from. It may also be given as a general direction, that as few stations as possible are to be used in filling in the detail of a survey—a station from which several lines radiate being more correctly determined than if only a single line commence or terminate at it. It might not be here out of place to mention the subject of *marking* stations in the field, in order that they may be readily referred to when required; and, as it frequently happens that months intervene between the putting down of a station and the reference thereto—as in the first triangulation, and subsequent filling in of a survey, it is most essential that the spot should be so marked as to place its site beyond a doubt. It is customary with careful surveyors to drive in a picket or stake at every station;—which is a custom that should invariably be pursued—but as mischief or accident will sometimes effect the removal of such pickets, it is necessary to define the spot by paring the turf round it of a triangular or of a star shape, thus: or, if the station should be in a ploughed field, then either dig out the ground, or indent it a similar shape with your heel. Where stations are required in roads or on hard stony ground where stumps cannot be driven, it will be necessary to drive in large spike nails, or to scrape a hollow, and fill it up with stones; at the same time, to prevent doubts, a scrap of paper with the chainage written thereon may be covered up with the stones; many months after stations similarly marked have been formed, we have referred to them without difficulty. In surveying a town, great difficulty will be often experienced in marking stations, and in subsequent reference thereto; and, if the streets in which stations are necessary should be pitched or paved, it will generally be found impossible to mark the precise spot. In our practice, when such is the case, we drive a spike nail to the right or left of the station, as may be convenient, and enter in the field-book the lateral distance or eccentricity. If we have occasion subsequently to refer to the station, no difficulty is experienced in deciding on the exact spot. Before concluding our remarks on stations, we may mention that small pieces of tin for station marks will be found greatly superior to the common "whites" mentioned at page 8, the rays of light reflected therefrom rendering the tin marks conspicuous, when the papers are barely visible.

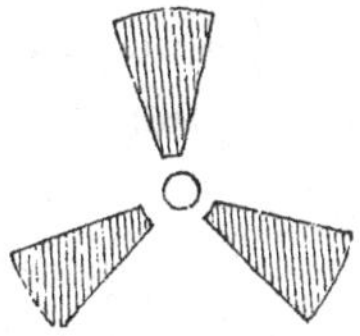

The length to which offsets should be limited is a matter not easily to be determined on. The tithe commissioners, in their plans, have limited offsets to one chain, but, in certain positions, offsets of two or three chains may be taken with great safety; and, again, in other positions, offsets of even one chain are far too much. In the first instance, if a line is being measured by the side of a fence, which is nearly straight, it will signify but little if the offsets exceed a chain; and it will often happen that much time and trouble will be saved by taking an offset of two or three chains to a junction of fences, but, in such cases, the angle formed by the offset with the main line should be taken; if an angular instrument be used, the offset may be taken from any point on the line without regard to its being a right angle: or any point in the fence may be determined by throwing out a small triangle, as in the diagram at page 35; but if it is required to determine the position of the fence by *one* direct measurement from the main line, and no instrument but the chain is at hand, it will be advisable to take the offset at right angles, as well for convenience of measurement as plotting. But as an offset of a chain in length cannot be correctly set off at right angles with the eye only, we will point out a correct method of setting off a right angle by simple measurement: suppose it is necessary to measure the line b c at right angles to a b, from the point b;—to do so, set up a mark at any convenient spot d, and measure the length b d;—measure out the same length d b, from d, touching the line b a, at a; then retire to c, in a line with a d, and set up a mark at the exact distance from d that a was previously put down: c will then be at right angles to the line a b, at b. Another method for accomplishing the same object is to fasten

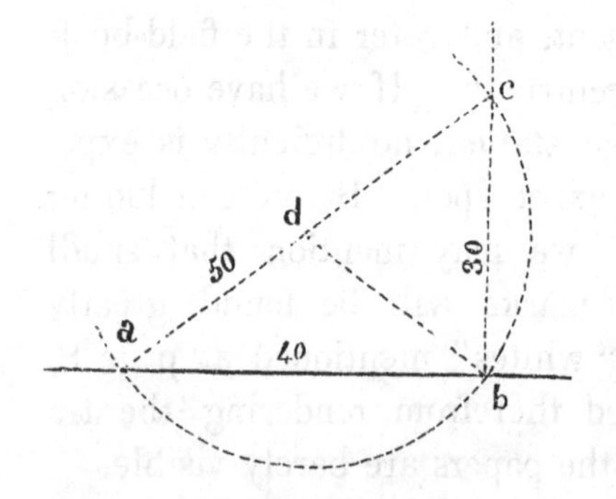

twenty links of the chain at b,—measure forty links from b to a, and there fix firmly the handle of the chain. One end of the chain being then fixed at a, and the other end—wanting twenty links—at b, there will necessarily remain eighty links of the chain between b and a: if the *centre* of the chain be then drawn tightly in the direction of c,—c b will measure thirty links, and c a, fifty links, which is the proportion of the sides of a right-angled triangle, *viz.*, three, four, and five*—the square of the hypothenuse, fifty, being then equal to the sum of the squares of the other two sides, forty and thirty; the line b c is consequently at right angles to the line b a. In "taking up" a fence or boundary, the number of offsets requisite, and the intervening distance between each offset, must of course depend on circumstances: first, as to the degree of accuracy required; secondly, the scale to which the survey is to be plotted; and, thirdly, the direction and straightness of the fence itself. In the first place, if a minute detailed plan is required, from which property may be safely sold or purchased, it will be necessary to take very frequent offsets, and the slightest bend which may exist will have to be shown,—if the direction of the fence is pretty uniform, either as regards a straight line, or in a gradual sweep, an offset at every chain will probably be sufficient—but, if the fence should be newly made and straight, an offset at either extremity, and one or two intermediate, will be sufficient; the latter being necessary to correct any mistake that may arise in the measuring of either one of the extreme offsets, and which would then be detected in plotting, from the extreme and intermediate points not lying in the same right line.

* Builders commonly make use of these proportions in setting out the rectangular walls of houses; but any equal multiples of those numbers, as six, eight, and ten, may be used. Draughtsmen sometimes adopt the same mode in setting off rectangular lines on paper, if a square is not at hand.

In the second case, if the scale of the survey is minute,—such as five or six chains to the inch,—the offsets need not be so numerous as when the scale is of a larger size. A small difference in the distance of the fence, which would be quite visible when plotting at two or three chains to the inch, would be quite inappreciable on the smaller scale. The third case, which relates to the degree of irregularity of the fence itself, is of course the main cause of many offsets being required; but, by judicious management, even in the most irregular fences, the number of offsets may be greatly reduced from what at first sight appears necessary. For instance, take a serpentine brook, —we know that a stream of water will pursue as direct a course to its outfall as the nature and course of its channel will permit, and, therefore, the continual and violent abrasion of the irregular or angular parts of its channel, until it either becomes straight, or of such *regular* curvature as to offer little resistance to the flow of water; and which will generally be found to be the case in the deflections of all rapid currents. Angular deflections of hedge rows are also generally uncommon, from occupiers of land finding it to their interest, in farming, to obtain as straight fences as possible; consequently, crooked angular fences are as speedily reduced into curved lines as circumstances will permit. By simple attention to this fact, the labour and time generally consumed in taking up a crooked boundary, may be vastly economized; for, by taking offsets to the centre and extremities of a curved line, the intermediate portions can be easily and correctly drawn on the plan. The preceding enumerated methods of determining offsets will, however, not always be available, as in the case of part of a fence setting back two or three chains from the general direction, when it will be found the most advantageous plan to throw out a small triangle or trapezium, to offset from; as explained in the accompanying diagram:—The line a b is the chain line, from which the

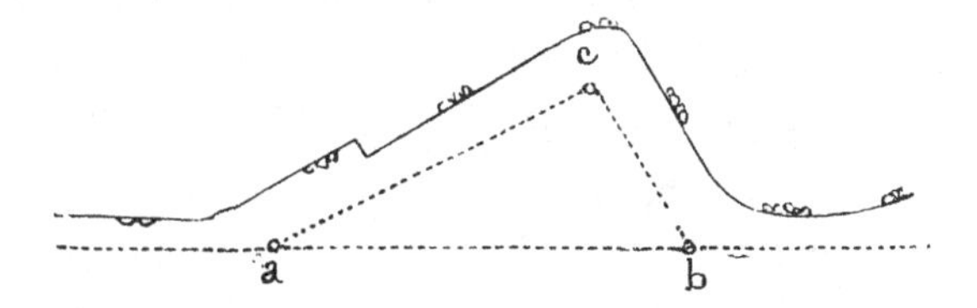

position of the fence is determined by offsets; but a sudden set back at c from the general direction, renders it impossible to offset to that part correctly, therefore the small triangle a c b is thrown out, and from the sides of which the offsets to the fence are taken. Offsets to all buildings, where the breadth as well as the distance from the chain line is required, should be carried on continuous: thus, an offset of twenty links to the side of a building which is twenty links wide, should be entered 20 "to" 40, and so on, for as many dimensions as are required, adding the word "to" between the several entries. A saving of time, and greater accuracy, would be effected, if this method was generally adopted. The shape of all buildings and small enclosures should be sketched in the field-book before the dimensions are taken.

We have not yet spoken of the varieties of fences which divide property, and how the surveyor is to determine the precise boundary line. Now, when it is considered that in general the actual fence is not the line of division, and that the precise boundary is only to be traced by a careful observation of the actual fence, and knowledge of the relative position under similar circumstances, we think a few general directions will be acceptable. Where a common ditch and hedge, or a ditch and bank, divide property, the brow, or inner edge of the ditch, is generally the boundary; but in many localities, where a ditch and hedge is employed to define boundaries, the actual line of division is not the brow of the ditch, but the foot of the bank, and more commonly the root of the quicks, or, as it is termed, "the stake."* The centre of "a balk" dividing "common lands" is generally considered the boundary, and also of an open ditch, water course, or dyke. The width generally allowed for a ditch is six links, and for a bank or hedge nine, making fifteen for the ditch and hedge together; it is, however, absolutely necessary in these matters that the surveyor make inquiries as to local custom in every instance. Where a boarded or post and rail fence occurs, coupled with hedge and ditch, it will often be troublesome to ascertain the precise boundary of the property; but a guide will be found, by observing from which side the nails are driven—it being generally understood that nails are driven homewards; if driven on the ditch side, the brow of it will be the boundary; and if on the other side, the wood fence itself.

In lanes and parish roads it is very common for the ditch to be on the field side, and the bank and hedge towards the road, without any ditch on the road side; and sometimes a ditch may be observed on both sides. In the former of these cases, if the road is made quite up to the foot of the bank, take that as the boundary; but if not, and any portion of waste intervenes between the road and the bank, allow for the width of a ditch.†

* In parts of the country where much wood land exists, "a brow," or narrow strip of underwood, is a common boundary, without ditch, bank, or hedge, to define the precise line of demarcation. In such cases, the only rule to be observed is, to offset to the oldest stubb in the brow, or to any timber trees or pollards that may be standing within it; should none such be observed, offset to the centre. In one or two instances I have observed a curious line of division between property, the origin of which I am at a loss to account for. It is termed a rod-fall, *i. e.*, a rod over the hedge, from the brow of the ditch.

† The general application of this last direction is doubtful, as in many parts

In the latter case, where there is a ditch on both sides, take the road side ditch as the boundary, the other being made for the purpose of drainage.

Where a fence changes from one field to another, *i. e.*, the hedge lying in a different direction, it must be carefully noted in the field-book, and the breadth of the fence set back on the plan, as in the diagram at page 35. In every instance stiles and footpaths must be laid down on the plan, and many surveyors insist on the propriety of the gates or openings to every enclosure being entered.

Surveyors frequently adopt reprehensible practices in measuring right lines; among the foremost of which we may mention that of using only nine pins or arrows instead of ten, making a huge mark or hole in the ground in place of the tenth. In measuring the next chain forward, the handle in such case is held over the hole—generally a link broad—instead of up to a pin, often thereby introducing considerable errors, if the line is of any great extent. We consider the best method of changing the pins is, for the follower to give one to the leader on the tenth being put down; then the next pin will be the eleventh, or the first of the new series; but before the leader draws on the next chain, the follower advances to him, and delivers up the nine pins remaining in his hands, when the chainage proceeds in the regular order. Another method in common practice on level ground, is, to measure out the eleventh chain, and carefully lay down the handle when in correct position; the follower then advances with the ten pins, and fixing one at the extremity of the chain, gives the remaining nine to the leader, thereby effecting the change. Another method is, by making use of eleven pins, the eleventh being shorter or different

of the country I have observed in such cases that the foot of the bank was considered the boundary.

in some respect from the others, and used for marking the eleventh chain; the follower substituting another pin for it before the leader draws on the twelfth. The only further remark we have to make on this subject is, to insist on the necessity of every change of pins being entered in the field-book; if otherwise, where enclosures are large, there is the probability of an error of ten chains being made on some of the lines.

It will frequently happen, on large surveys, that some of the lines for taking up fences require to be measured along the middle of a hedge for some distance,—or a station formed, and a new line commenced from the point where it touches the hedge,—or the line altogether given up, and a new line set out clear of such obstacle. It is troublesome to adopt any one of these methods, and we conceive a simple plan for avoiding a hedge in such cases, without abandoning the line, will be serviceable. It may be effected thus:—a b is the chain line passing

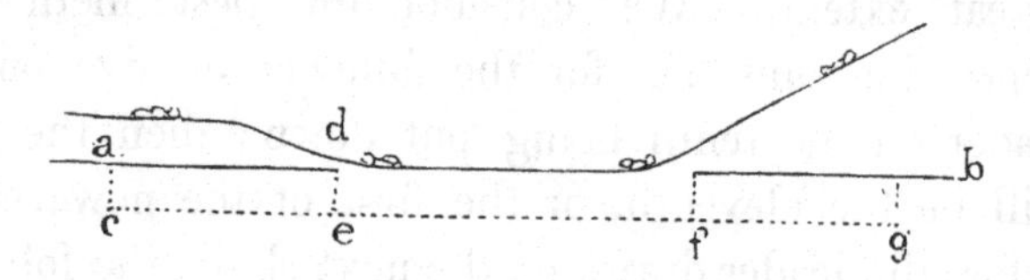

into the hedge at d; at any convenient distance from d, as a, set off a short measurement, a c, at right angles,—say ten links, and also set off the same distance from d; a line, c e f g, parallel to the original line, can then be measured for any required length. Of course, the greater the distance between c and e the more accurate will be the process; and, when quite clear of the obstacle, the same method pursued on the reverse side will remove the line into the position it would have occupied had the measurement been continued forward without interruption.

Another difficulty which often occurs is, the chain terminating in the middle of a hedge, brook, or pond: in such case, move the preceding pin forward, ten, twenty, thirty, or more links, until the length of the chain from such pin will reach beyond the obstruction, taking care to make the necessary deduction from the next chain's length.

In laying out a line for the purpose of taking up a fence which has considerable angular deflections, it may often appear necessary to have elbows in the line, or, in other words, "to form a station," before closing on any determined point, and then go off in another direction, for the purpose of taking up the remaining portion of the fence. But such a system should never be adopted, not even when angular instruments of the best construction are employed, as lines intersecting at very obtuse angles—such as the deflections of fences generally present—require such extreme nicety, both as regards the measurement of distance and direction, as to preclude the chance of correctness in common operations; and it should be borne in mind, that, in the generality of cases in surveying operations, in proportion as the difficulty of obtaining correct results increase, should the cases involving such difficulty be avoided. The annexed diagram will more clearly explain what we have just remarked on this subject.—The line a b c, we may first remark, is so laid out as to take nearly a *mean* course in regard to the fence—which is an advisable method to be pursued if it can be done without great inconvenience — by

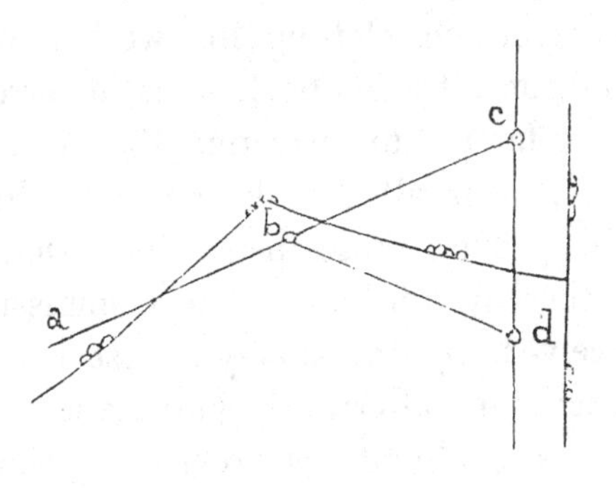

which means a fence much deflected from a right line may often be brought wholly within reasonable offset distance; but if this last-mentioned consequence should not follow, and the fence so suddenly deflect—as at b in the diagram—that the remaining part of the fence is without offset distance, the line must by no means be terminated at b, and a new line b d taken, but a station must be formed at b, and the line continued on to c; after which a line from b to d is to be measured, to take up the remaining part of the fence. We need scarcely remark that were the lines a b and b d alone measured,—the extension b c being omitted,—the most trivial error in either of the distances would so much alter the position of b, as wholly to change the appearance of such part of the survey, and of course produce a considerable error in the separate area of the two enclosures.

It has been a matter of considerable discussion, whether field-books should be kept in ink or pencil. The Tithe Commissioners, in their instructions to surveyors, insist on the former method; and we believe such have also been the directions to the persons employed on the Ordnance surveys. The chief reason for insisting on ink registers for the tithe surveys appears to have arisen from a want of confidence in the persons employed, and a fear that the surveyor, on plotting his work, would, if he discovered an error, alter his book so as to correspond with the map, in preference to correcting the error itself. But we conceive, with all due deference to the talented individual who penned those instructions, that no security whatever is thereby afforded. The commissioners may be easily deceived, by the surveyor first keeping his field-book in pencil, and afterwards putting it in ink, and making the *necessary* alterations previous to remitting it to the Tithe Office. Or an *erasure* from a field-book kept in ink—

where it would not bear comparison with the map, would quite defeat the original intention. A strong objection with us to keeping a field-book in ink, is, the utter impossibility of working in wet or misty weather, from the ink blotting on the paper, as well as the unpleasantness of erasing when a wrong entry is made. If it is considered an essential, that an entry once made in the field-book shall not be expunged, it may be effected by using a prepared paper, highly glazed, with a metallic point for writing, and which is immoveable, without any of the disadvantages attendant on entries in ink. Field-books of asses' skin are commonly employed where surveys for railways and similar works have to be completed within a limited time, regardless of bad weather. Respecting the several forms in which field-books have been kept, we shall speak of such in the next chapter. It may be proper in this place to mention that, where several persons are engaged on one survey, the same system should be rigidly adhered to, so that in case of casualty one person may plot another's work.

All maps and plans should be constructed with the north upwards, and of course the west to the left; have a title, scale, and reference, with the date of performance, and other necessary particulars, neatly written thereon.

In concluding these preliminary remarks, we have only to observe that no positive rule can be laid down for adoption in every case that may occur in surveying. The reader should make himself acquainted with the best general methods, and exercise his own discretion in the matter.

CHAPTER IV.

EXAMPLES IN SURVEYING.—SURVEY OF THREE FIELDS, WITH THE FIELD-BOOKS.—SURVEY OF A SMALL ESTATE, WITH FIELD-BOOK.—PARISH SURVEYING.—REMARKS AGAINST THE TITHE COMMISSIONERS' SYSTEM, AND PROPOSED IMPROVEMENTS.—LAYING DOWN PARISH SURVEYS.—SETTING OUT LINES, ETC.—GENERAL REMARKS.—EXAMPLES IN RAILWAY SURVEYING, WITH AND WITHOUT ANGULAR INSTRUMENTS, WITH FIELD-BOOKS, AND REMARKS.

WE now submit to the reader's notice a few practical examples in surveying with the chain alone, and also in combination with angular instruments. The three fields represented in plate 3 should be surveyed by the same method, whether an instrument was used or the chain only; but in the present instance we will confine ourselves to a description of the survey as effected without the aid of any other instrument than the chain.

In the first place, it may be observed, generally, that, before commencing a survey, the extent, shape, and position of the ground to be taken in, should be carefully examined, for the purpose of the more readily laying out the lines of construction to advantage. In the plan of the three fields referred to, we perceive that a line in the direction A B is the longest that can be obtained through the property; which will generally be found the most proper rule to adopt in laying out a base line, or diagonal, as it becomes when the trapezium A C B D is formed about it. We commence operations at A, a little without the boundary of the property, at which spot it will be necessary to put up a mark (such as we have already described and termed "a white"), and set out the line to B; cutting down the intervening hedges, without the mark at B, or an object in a line with it, should be visible. When the line is ranged out to the desired extent, commence the measurement of it from A towards B; but if the mark taken at B, or in a line with it, can be seen at every part on the line A B, then there will be no necessity to *range* out the line beforehand, but the measurement may be proceeded with at once. The measurement being then commenced, is proceeded with, without remark or entry in the field-book (except the crossing of the hedge near to A), until the arrival at a, where a false station is left, and the measurement referring to the same entered in the field-book; as also the crossing of the hedge, the false station at b, the crossing of the next hedge, and the termination of the line at B, which completes the base line, and on which the position of the other lines depend; although, as far as the actual surveying or delineating of the fields is concerned, no use is made of the base line (but where fences run nearly parallel with it they should always be taken up by offsets), yet the intelligent reader will at once see its importance, and that any error committed on it will affect all the lines used in the survey.

After completing the measurement of the base line up to B, a side line, B D, is to be measured,—the line being ranged out beforehand if an object at D cannot be seen from every part of the line;—but with skilful surveyors side lines are rarely, if ever, set out before measurement, as a mark that can be taken will generally be found either at D, or beyond it at y, or, if not, on nearly deciding the direction of the line, a mark, backwards at x, will generally be found; the line B D can then be measured without inconvenience or trouble, by backing it, *i. e.*, the surveyor—if his assistant cannot be depended on—takes the leader's end of the chain, and places his pin exactly in a line, with the starting point and the back mark, which process is generally repeated until a forward mark in the same line is discovered; if such should not be the case, occasional marks are set up as the line advances, which guide the leader in keeping his line, or "plumbing back." In the measurement of this line, offsets are to be taken wherever necessary, and the crossing of the hedge, and false station at c (the purport of which will be understood on inspection of the plan), entered in the book. On arriving at or about D, it is probable the mark left at the first station A will be visible, which, if such is the case, measure to it (taking offsets, and leaving a false station at d); but it will more often happen than otherwise that the station at the commencement will not be visible from D; if so, its proximity must be ascertained, and a line measured to an object a little within it, so as to insure the intersection of the line A B with the line D A, as, if this be not attended to, and the line be measured so as to fall without A, it will be necessary to *back* the line A B to z, or to such distance as the lines will intersect; which quantity from A to z must be added to the distance of the stations, &c., on the line A B, and the point z will then become the starting point. If the line D A should fall within the point A,—as at w, for instance,—the distance from A must be ascertained, and entered in the field-book, as well as the distance from D to the point of intersection, which will be entered in the field as *a close*.

Commence again from A, and measure a line to some object at C, of course noting the offsets, crossing of the hedges, &c. When arrived about e, put down a mark in a line with the two false stations previously put down at b and c, and continue the measurement on to C, at which point the station mark at B will probably be visible, up to which measure a line from C, which will complete a great trapezium, enclosing the property; but, if the station at B should not be visible from C, the same method must be pursued as directed for the measurement of the line D A.

The detail of the survey now remains to be filled in; to do which, first measure the secondary line e b c, which is straight, and cuts the base at b; when this last line is measured as far as f (which is in a line with the two false stations at a and d), put down a mark, and note the particulars in the field-book, also the exact distance from e to b; continue on the chainage to c, which will be a close. Then measure the line f a d, carefully noting the chainage on arrival at each false station. These internal lines, it is evident, will form a check on the measurement of the triangles A D B and A C B.

The field-book or register of the above survey is given in two separate forms, as will be seen on reference to plates 1 and 2. The first method—which is the most common—is merely a single line ruled up the middle of a page of an oblong field-book, which form of book is much more facile in the field than when of a square shape. Upon this line, which represents that under measurement, all the dimensions are written, the offsets, as in the former case, being entered on either side. The chief difference between this field-book and that given at page 11 is in omitting the double line for the direct distances, and in referring to the *lines* by numbers, instead of to the *stations*, by letters. In very extensive surveys, such as a parish of some thousand acres, the main lines, and in some cases the principal and secondary lines, are referred to by letters, and all the other lines by numbers. We have occasionally adopted a plan in our field-book for facilitating the reference to lines, which we think may prove generally useful; it is simply to *page* the field-book, by which means a reference can be made to any station, on any line, with the greatest facility. We do not consider it necessary to give a detailed description of this form of field-book, and shall content ourselves by merely referring the reader to it; but we would beg to throw out a hint for his guidance in the matter, which is, to make himself thoroughly acquainted with the detail of the register, and plot the survey several times from it, without reference to the map, whereby he will not only acquire great facility in reading the field-book, but the principles of the science will be gradually instilled into his mind; indeed, we know of no better method, in the first instance, of acquiring a knowledge of surveying than by plotting a map from the field-book.

The second form of field-book, and which is represented at plate 2, is an eye sketch of the survey, made as the work proceeded. It will be perceived that each line and fence occupies somewhat like its true relative position; and certainly, if nothing more can be said in its favour, than that it is simple, requires no reference to other parts of the book, nor any effort of memory in plotting, and, withal, can be perfectly understood by any stranger, even such as are of the most common capacity, it must be admitted that some benefits would attend the use of it, and authorize its more general adoption. But there are other advantages attendant on its use, such as a more rapid execution of the field register, and greater facility in plotting, from the relative position of the lines and the whole of the dimensions being embraced in one view. The only explanation we think required in regard to this form of field-book is, that the direction of the figures represent the lines under measurement, and the dimensions are entered from the commencement, upwards,

FIELD BOOK TO THE SURVEY. IN PLATE4.

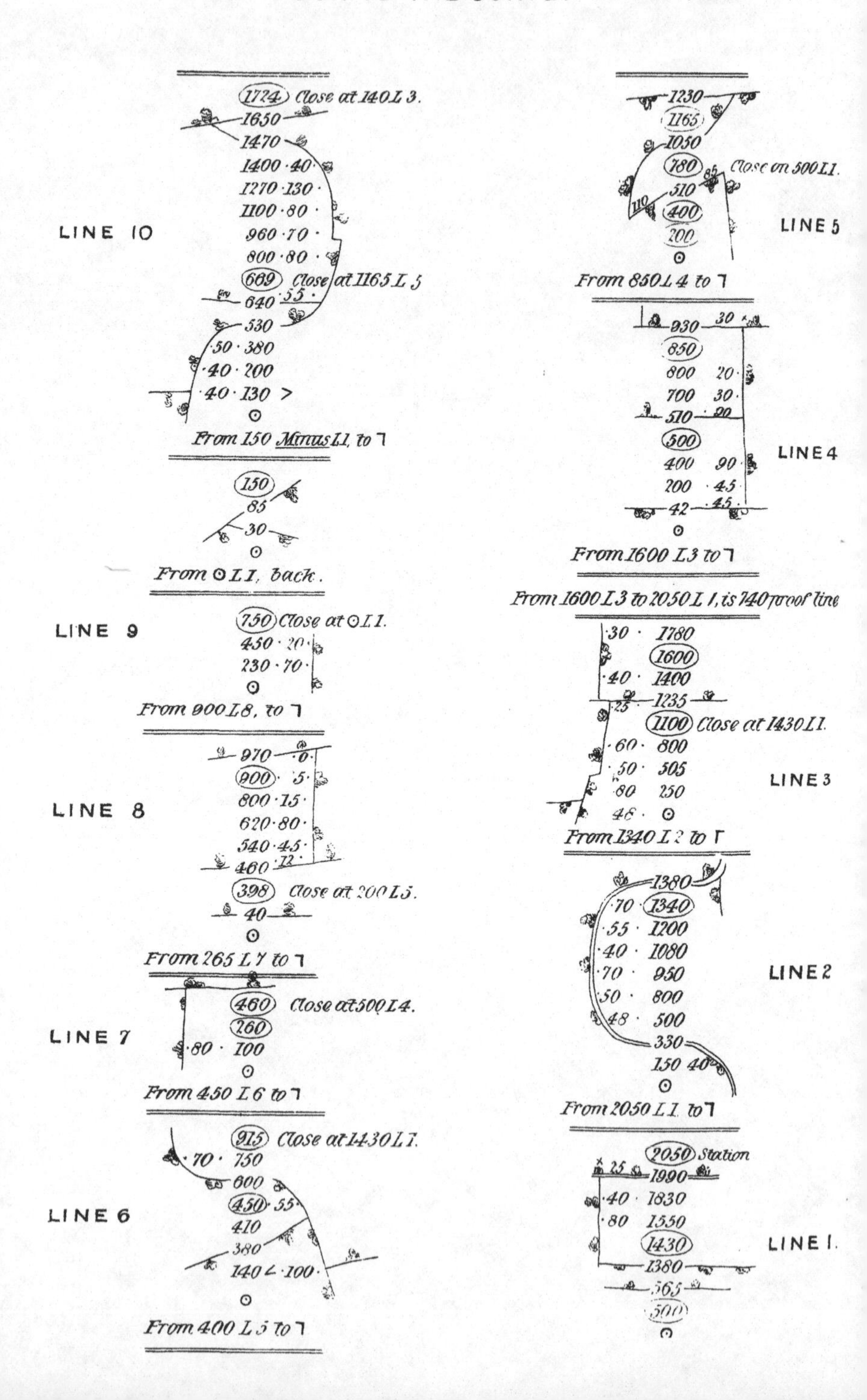

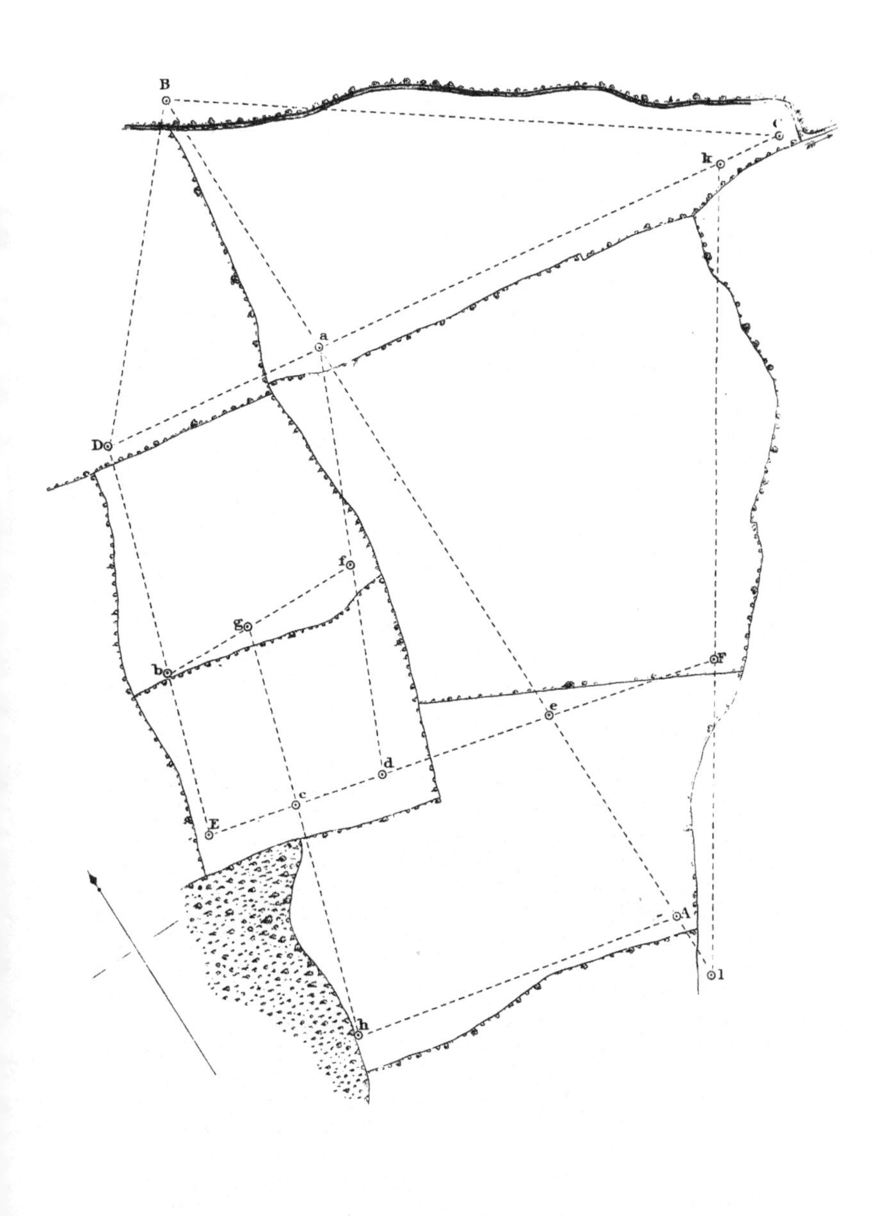
B
C
k
a
D
f
g
b
F
e
d
c
E
A
l
h
200 Links
0
5
10 Chains
SCALE

exactly in the same manner as in the form of field-book with the single vertical line.

It may be considered by many readers that the sketch form of field-book must be confined to surveys of very limited extent, but we can assure them that with a little practice it may be successfully used for surveys of any magnitude. The description of book necessary to be used when the sketch register is adopted, must not be oblong, but nearly square when open. We generally find the small cyphering books, with stiff covers, commonly used in schools, to be the best suited for the purpose. The entries should be made clear and distinct, with a hard sharp-pointed pencil, when there will be little fear of their becoming illegible; or, if such should be feared, our suggestions on the subject in the last chapter might be adopted.

The survey represented in plate 5 is somewhat different in its character from the preceding example, inasmuch as the enclosures are more irregularly disposed, and consequently require some modification in the arrangement of the lines. A B is the base line,—which, as we have previously remarked, as a general rule—is carried through the greatest extent of property to be surveyed. After the measurement of the base line was finished up to the point B—to which point it was continued, to allow a mean line to be laid out for the purpose of taking up the adjoining fence—the line B C was measured. From C an object at D was taken, to which a line was measured, crossing the base at a, at or near to which point a station was left when measuring the base; the junction of these lines being noted in the field-book, the measurements entered, &c., the line was continued to D, which was considered a favourable point for commencing the line D E. At this part of a survey so circumstanced, it would be prudent to measure a tie line across from D to B, to check the relative positions of those points, although it would not be necessary, could accuracy be insured in the measurement of the sides of the triangle, a B C;—a D, it will be perceived, being an extension of one of those sides. But to return to D; from this point was measured the line D E, leaving an intermediate station, b. From E was measured the line E F, leaving intermediate stations at c and d, and crossing the base at e, at or about which point a station was left when measuring the base. The junction of these lines being effected, the measurement was continued on to about F, where a *temporary* station was formed. We then returned to the station left at d, and measured a line to the station at a—if it was visible, or, if not, to the nearest point where it was supposed to be. When measuring this line at about f, the station left at b was seen; but, if such had not been the case, the station f would have been left at any point considered most advantageous, and the measurement then carried on to a, which was a close. We returned to f, from which we measured a line to b, or to the nearest point where we expected to find it. On measuring this line the station at c was visible, and the intermediate station g was put down so as to form a direct line with the station at c and some natural object at h; but, in similar cases, where no natural object exists at h that can be taken for a mark, it is probable such an object may be observed at i; the station g may then be put down in a line between c and such object. The line g h was then measured, the junction at c being noted. From h, a line to A—the mark left at the commencement of the base line—was measured, which closed the work on that side the base. The line l k, however, remained to be measured, to effect which, it was necessary to *back* the base line from A to about l, of course noting the distance. When this point was determined on, the line l k was measured, passing near to the *temporary* station left at F; the measurements E F and l F, to the point of juncture, being noted, and in like manner the close of the line l k on C D, presuming the former line not to have been measured direct to the station left at or about k. We have here, in a detailed manner, shown how this survey was effected, but have yet a remark or two to make thereon. It will be seen, on looking over the plan, that the principal lines which we have described as being measured, are all disposed in triangles, and the minor lines, used in filling in, either form similar figures, or trapeziums, having the angular points fixed by the previous mentioned triangles; and in no case is a point fixed by the intersection of two lines only; for in every instance it will be seen that, where a trapezium is formed, three sides are fixed by the previous triangulation; and, where the figure is a triangle, that the position of two sides is checked by a third, in addition to the remaining side. Thus, the point D is determined from being on an extended part of the line C a, forming one side of a triangle; and is also checked by the measurement D B. The point E is also fixed, from being on an extended part of the line e d, forming one side of a triangle, and the measurement of the line D E becomes a check on the accuracy of the previous operations in fixing those points. In like manner the point h is determined by the measurements g h and h A, and also from the position of the line g h being determined by passing through the two fixed points, g and c. The position of C on the other side of the base is certain to be correct, from the point D being proved so. We think no further explanation is required relative to this survey, and therefore refer the reader to the field-book for any further information that he may require.

Parish surveys being usually performed with the chain alone, we think, in continuation of our subject, we cannot do better than introduce it in this place; but, previous to making any remarks of our own, or giving any directions on the subject, we think it best to give some extracts from Captain Dawson's Report to the Tithe Commissioners, relative to the method in which those surveys are to be effected, premising that such method is imperatively necessary where the chain is used alone, but when combined with a good theodolite, the necessity for a strict observance of the instructions referred to becomes questionable. "It is necessary," Captain Dawson observes, "to determine the area of the whole parish by some means, which make the correctness of that area independent of the result obtained by summing up the contents of each enclosure, minute errors in many of which would escape observation, if not checked by comparison with the correctly ascertained whole. It is essential, in fact, to arrive at the total area of the parish by direct admeasurement of the space included within its external boundary; and the simplest and cheapest means by which a survey and plan may be made for effecting this object appear to me to be as follows:—

"First,—To measure two straight lines through the entire length and breadth of the parish.

"Secondly,—To connect the ends of these lines by means of other measured lines; and,

"Thirdly,—From these connecting lines (by measured triangles and offsets) to determine the entire parish boundary.

"The true area of the parish may then be obtained by calculation from the measured distances, and by the admeasurement of the included space upon the plan. Lines of the description herein proposed to be measured are ordinarily used by surveyors in the construction of their plan, but are not always shown on the finished map; I propose to retain them permanently for purposes which will presently appear. The object and application of these lines will be better seen by reference to the accompanying diagram, which is a rough sketch of a parish to be surveyed.

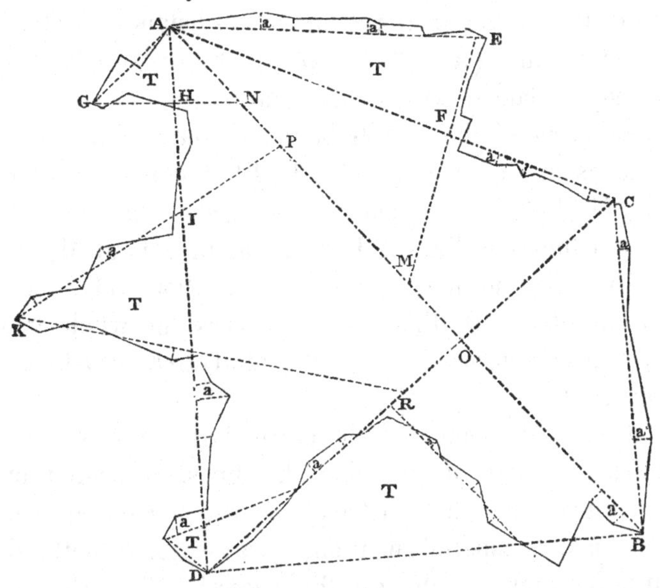

The two main lines which I should recommend to be measured through it are marked A B and C D; A C, C B, B D, D A, are the connecting lines. a a a are the offsets, or perpendicular distances of the several angular points of the parish boundary from the measured lines. Now, if the main lines A B and C D be measured accurately, and their true lengths from the point (O), at which they cross one another, be laid down upon the plan, it will be seen that the connecting lines A C, C B, &c., will form an efficient check on the general direction of the two main lines with reference to one another. A satisfactory check on the *lengths* of the several lines will, by the same means be afforded; for, as the points A, C, B, D, are in each case determined by the intersections of three lines, an error in any one of these lines must immediately be discovered. Thus the true relative position of four extreme points (A, C, B, D), in the parish boundary, will be obtained, and such portions of the boundary as fall within the ordinary range of offset-distances from the connecting lines (A C, C B, &c.) will also be determined, and may be laid down in their true positions. The more remote parts of the parish boundary may be determined by means of the triangles (T, T, T), the sides of which (E F, G H, K I, &c.) being prolonged on the ground to intersect the main lines A B, C D (as they do at M, N, P, &c.), may be laid down correctly in position and direction upon the plan. By this simple process the whole boundary will be determined, and the total area may then be ascertained.

"Among the objects to be particularly attended to in practice, is that of reducing the lines, measured over steep slopes in hilly districts, to the horizontal plane. This demands especial mention, because some inattention to it is not unusual, though the necessity for such reduction is well known to practised surveyors, and all should be alive to the importance of using a theodolite, spirit-level, or other assured means, in the measurement of lines over hilly ground, for determining the exact allowance to be made. Without this reduction of the lines they cannot be laid down in plan upon a flat surface, and distortion of the outline must inevitably result. Care, of course, must be taken in all cases to measure the lines straight to the points desired; and this will require more particular care in a mountainous, rocky, marshy, wooded, or thickly inhabited country. The expedients in use among practical surveyors will of course be resorted to for overcoming any difficulties which may attend the measurement of these main lines, and the theodolite offers a never-failing resource in all cases where a departure from the direct line is inevitable.

"The main lines should be selected, as much as possible, with reference to permanent well-defined objects, such as churches, &c. In other cases it will be desirable that the extremities of the lines (or of some of them at least) should be marked, and preserved on the ground by stones or posts, or by trees, planted there so as to admit of the points being referred to at a future time. The lines which have been described as essential to be surveyed should, in all cases, be marked upon the plans. They should be drawn in red ink, in order to distinguish them from, and prevent their interfering with the lines of fences, &c.; and the length of each line in links should be marked in red figures upon it. In all cases of fences, the actual boundary line of the adjacent properties should be marked upon the plan, whether it be the central line, or the side of a hedge, ditch, wall, bank, &c.; and when the fence belongs entirely to one property or the other, that should be indicated by the proper mark. The parish boundary should be shown, in all cases, by a dotted line; and when it passes along the middle of a fence, the dots should be drawn on both sides of the fence, thus:—

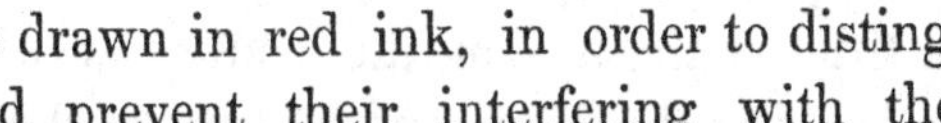

PARISH BOUNDARY

FENCE

"When a road forms part of the boundary of a parish, both fences of the road should be shown; and it will be desirable also to mark the abutments of other fences upon the outer fence of the road. The same remark will apply to rivers generally; and in Lincolnshire and other fen districts, to droves and the drains by which they are bounded, &c. When a parish boundary passes through a field or other enclosure, without being defined by a fence, the whole of such field or enclosure should be shown on the plan, with the parish boundary, marked by a dotted line, passing through it. The area of the included portion only of such field or enclosure will appear in the schedule; but the area of the excluded portion may with propriety be given on the plan, and be marked as belonging to the adjoining parish.

"The plans are to be drawn to the scale of three chains to one inch, to admit of the correct computation of the contents of the several lands. And the ordinary usage should be observed with regard to placing the north towards the top of the plan, writing the name of the parish, as a title, with that of the county in which it is situated; and adding the name and address of the surveyor, the date of performance, the scale, and the total contents."

The instructions to the tithe surveyor contained in the foregoing extracts are unobjectionable where the parish to be surveyed is nearly of a square shape; but, where the boundary forms an oval, or long parallelogramic figure, the method is decidedly objectionable; and the objections increase where, as is often the case, the parish is oblong, narrow in the middle, and swelling out in width at the extremities. Our chief objection to Captain Dawson's *universal* system of surveying is, that, without the parish be nearly square, the two diagonal main lines must intersect obliquely; or, if laid out to cross each other at right angles, or nearly so, an immense waste of labour must result from it. In many such cases, where the system has been indiscriminately followed, we have no doubt double the area has been enclosed more than was necessary; and, as Captain Dawson estimates the triangulation at from 1*d.* to 3*d.* per acre,—and if we take the larger sum, which in ordinary cases is the least at which it can be done for,—we perceive how large the loss would be on a survey of very moderate extent; and, as this kind of surveying is very inadequately paid for, and no allowance whatever made for triangulation without the bounds of the parish, we leave the reader to please himself whether he will adopt the method under such circumstances. If, on the other hand, in following this system the surveyor keeps his lines of construction strictly *within* the boundary—the shape of the parish being oblong—the two main lines will intersect obliquely, and give rise to error and great uncertainty, which we have had occasion to speak of in a previous part of this work.

Before we proceed to the details of parish surveying, we will illustrate our remarks relative to the laying out the lines of construction, by means of some diagrams, each representing a parish so circumstanced as to require the rejection of the system laid down by Captain Dawson. The accompanying diagram represents a parish with the lines of construction laid down on it.—A B is the main

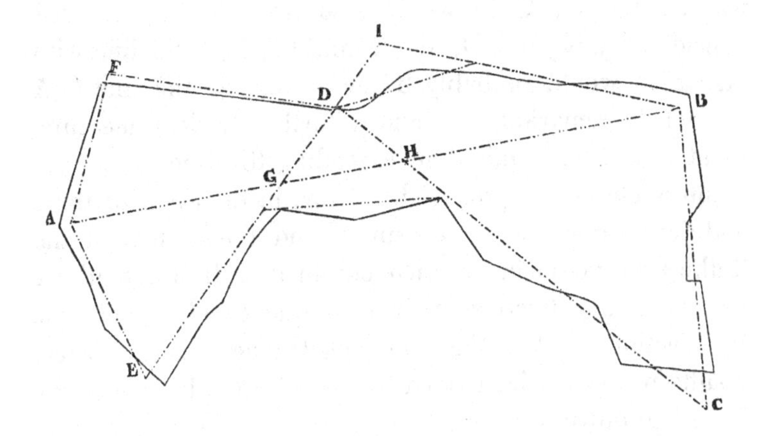

line or base of the survey; and B C H a principal triangle, the vertex C being determined by intersection, and its position checked by C H being produced to D. In like manner, the position of the vertex E of the triangle A E G is checked, by producing E G to D and I. The position of the vertex F of the triangle A F D—the points A and D being fixed, the distance from A to D is considered as base to the triangle—is, however, not checked; but, in filling in the detail, many check lines must necessarily occur to prove its correctness. But if it is considered essential that the main lines check each other in every instance, then the line C D might be produced beyond D, until it intersects with A F produced; or B I might be produced beyond I, until it intersects in a similar manner with A F produced. On examining this diagram, and considering the system laid down by Captain Dawson applied to it, it will be seen how much additional labour would be incurred; for, supposing the lines A B and C D to represent the main lines running across a parish, then a line, C A, according to his plan, would have to be measured, which would be wholly useless in so far as the real work of the survey was concerned; C D would also have to be extended considerably beyond D, and connected by a tie line with A, which would probably be as useless as the line C A previously remarked on; and a further useless measurement would be required in extending the line B I, until it intersected C D, produced. Now, in the event of these additional lines being measured, and the system of the Tithe Commissioners carried out in its fullest extent, we unhesitatingly assert that, in consequence of the oblique intersection of the lines of construction, such correct results would not be arrived at, as by the plan which we have laid down.

A different system from what we have recommended may be adopted where the boundary of a parish forms a figure similar to the accompanying diagram. In such a case, probably the most advisable method would be, to modify Captain Dawson's instructions, and form *two great trapeziums* in the parish instead of one, with a base line measured through the greatest extent, forming a diagonal to both trapeziums.—Thus, A B is the base line, E F and

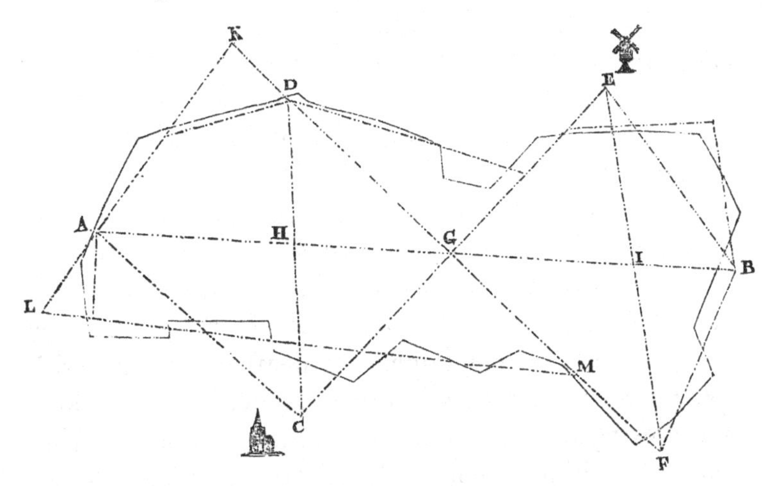

D C the transverse main lines. The tie line F G—forming one side of the trapezium, E B F G, it will be perceived, is extended to D and K, forming a side of the other trapezium, G C A K; and, in like manner, another side, E G, of one trapezium, is extended and forms a side, G C, of the other trapezium; then, the measurement of the main lines, A B, E F, and C D, will most effectually test the accuracy of position of the stations, C, D, E, F; and a further check will be afforded by the tie lines, K A C, E B F, and L M. The base line A B might be altogether omitted; or, at least, it might only be set out, but not measured, except the portions A H and B I, intersecting the main lines, C D and F E. Or, if it suits the operator's convenience better, the two last-mentioned lines might be omitted, retaining the base A B; then the measurement of F B would place the lines K F and A B in their correct relative positions, and would be checked by K A. In like manner A C would place C E in its correct position, and would be checked by E B. Some persons would probably wish for lines to be measured from C to F, and from K to E, as checks, but we consider, from the construction, that those points cannot occupy any other than their correct position.

The preceding remarks on the subject of laying out the lines of construction, together with the extracts which we have introduced from Captain Dawson's report, comprising as much information in this preliminary—but most important—matter as the reader can desire, we shall proceed to give directions relative to the measurement of those lines, and the entire completion of the survey; the reader bearing in mind that the instructions for parish surveying are strictly applicable to estates, and every kind of surveying, except for roads, canals, and other similar works, requiring a long narrow strip of country to be delineated, in which a totally different system must be pursued; but having introduced some examples in this chapter, we shall, for the present, dismiss the latter part of our subject.

The first thing to be attended to by the surveyor, on arriving at the scene of his labours, is, to inquire for any map of the parish which may be extant—presuming it to be extensive, otherwise it will not be worth troubling about; and procure the assistance of one or two intelligent men, well acquainted with the boundary and every object within the parish, to accompany him over it, so as to become conversant with its leading features; the most proper form of construction to be pursued, and the direction in which the main lines are to be laid out, will then be easily discovered. We have remarked, in our introductory chapters, that the lines of construction, or rather the base line—for in every system of operations the line *first* laid down on paper, and from which the position of the other lines are determined, must be the base or fundamental line—should pass near to the principal or most intricate parts of the survey; and we have now to remark, that the direction of the first, or of one of the other principal lines, should be laid down between two conspicuous objects without the bounds of the parish, as in the sketch at page 57, in which C E is represented as lying between a windmill and church. Where this plan is adopted, much labour is avoided in setting out the line in the first instance; and again, if at any time after completion of the survey it becomes necessary to lay out the lines on the ground, it is evident that, from the position of the line between the two objects being known, the other lines could be laid out with great facility; and, if the measurement of this principal line was commenced from one of the objects, the identity of the plan with the ground would be more complete. Where a line within a parish is measured in the direction of an object without the parish, such line is said to be *on* such object; the particulars of which should be written by the side of the line. In measuring the base lines of surveys for roads, &c., where they are *on* any conspicuous object, the particulars should be inserted on the plan, as at some future time much importance may be attached to the retracing and identifying the original base line. The following diagram is from Captain Dawson's instructions on the subject:— he observes, "that lines measured from any such objects (as Morden College) within the parish, in the direction of similar objects in the adjacent parishes—noting particu-

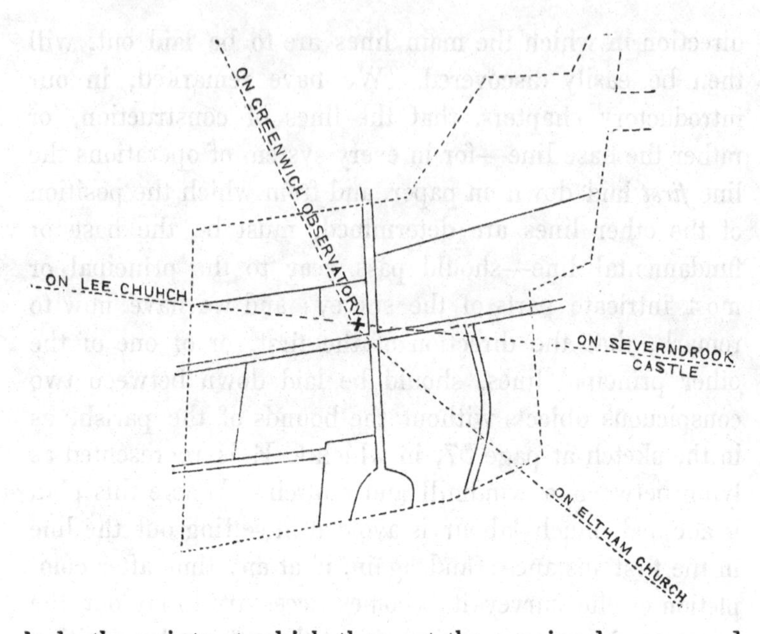

larly the points at which they cut the previously measured lines (*i. e.*, the outer tie lines), and the boundary of the parish—may be laid out upon the ground at any future time—so long as the churches and other objects remain, and the boundary be determined, though all the fences which now define it should be removed." The precise manner in which surveyors are to follow these instructions is not only very imperfectly detailed, but exceedingly apt to lead them into error. Whether it is intended that an object within the parish, such as Morden College, should be first fixed by triangulation and offsets in the usual manner, and then one *point* of it used as a station, from which lines are to be measured to other similar objects, we cannot tell; but, if so intended, it would occasion a great waste of labour, inasmuch as those lines would be of little use in filling in the detail. But if, on the other hand, it is *intended* that such an object should be fixed by the usual process of surveying, and then *angles* taken with a theodolite from it as a station, to objects without the parish, we can at once perceive the advantages which would result from it, and at a very small sacrifice of time.

The form in which the survey is to be constructed, and the direction of the main lines after due consideration being determined on, we will at once proceed to the measurement of one of those lines. It will rarely happen in ordinary surveying that one extremity of a line a mile or two in length can be seen from the other extremity, and therefore a difficulty arises as to the means to be adopted in such cases. We have in our introductory chapters described the manner in which a right line is measured from one object to another, when both are visible the one from the other; but in the event of a line, such as we have therein mentioned, being required to be carried beyond the mark towards which it was being measured, we have as yet given no directions. The importance of it in parish surveying, from the great length of some of the lines, is, however, great; and, before proceeding further, we shall explain the manner in which the difficulty has been usually overcome. A point does not determine the direction of a line, for any number of lines may have their source in the same point, without lying in the same plane;—two points are necessary to determine the direction of a line, and, when these are given, the line may be continued to infinity. These being the premises which we have to apply to practice, we are at once directed to a method of producing a line over ground of any kind of character: *viz.*, before losing sight of the first object, to set up a third; and, in like manner, before the second becomes invisible, to set up a fourth, and so on. The marks usually employed are common "whites," but some surveyors occasionally use poles of ten or twelve feet in length, with small flags on the top, but which method is much less accurate than the former, from the difficulty of setting up the poles perpendicular, and also in maintaining them in that position, which, if not carefully attended to, will occasion a departure from a right line. It is evident that, in setting out a line by this method on level ground, when the third mark is planted in correct position, the first will be hid by the second; and, in planting the fourth, the two first will be hid by the third; and how many marks soever are set up, they will all be *on* with, or rather hid by, the last, if it is planted correctly. Great care, therefore, becomes necessary in extending a line by a series of marks, and, to do it accurately, they should have no sensible breadth; but as, in that case, they would be imperceptible at any short distance, and consequently useless, it is proper to narrow them only within the limits of vision, but their depth may be increased to render them conspicuous. The top of the stick or pole on which the mark is placed is of course in the plane of the line, and it therefore becomes necessary that it should be plumb over the part fixed in the ground, otherwise the measurement will not be made in the vertical plane of direction. In setting out a line over an undulating country, as the preceding marks will not be hid by the last, but each appear vertically above or below it, great care must be taken to fix the marks in the vertical plane; and, if the ground rises or dips suddenly, it will be necessary to use a plumb line to test their verticality. The method of using a plumb line in such cases is this:—Put down a third mark as near as possible in a line with the two preceding, and advance some short distance beyond; holding the plumb line steadily in one hand, bring it into a position to cover the two first marks from sight; and, if the third is correctly put down, it will cover that also; if it does not do so, alter it until correctly placed; in like manner, however numerous the marks in sight may be, if correctly placed, the plumb line must cover them all. In this particular, the theodolite is eminently serviceable, and, when so applied, the operation is termed "boning." In tunneling, and in surveys where great nicety is required, "a transit instrument" is employed to trace out the line; but of this we shall have occasion to speak in another place. In the event of obstacles occurring to interrupt the continuity of the line, as a large tree standing in the way, the method given at page 38, for avoiding a hedge, might be successfully adopted. If a house, wood, or pond, should obstruct either the setting out or subsequent measurement, numerous available methods for overcoming the difficulty will be found on turning to a subsequent chapter. The line being then traced out, the measurement should commence from some fixed mark a little without the bounds of the survey. Great care must be observed in the measurement of a principal line, and the crossing of all fences and divisions of property noted in the book; but we would not advise false stations to be put down by the side of every fence, as is the usual method, but rather that marks be left at stated intervals on the line, for convenience of future reference. The reason for which we recommend this unusual method is, that, when filling in the detail by cross measurements from one principal line to another, it is scarcely possible to tell where the "close" will be; and, as it may frequently happen to be in the middle of a field, a great loss of time will take place, and some uncertainty arise, if, to determine the intersection, it becomes necessary to refer back to the station at the next hedge, and remeasure the main line from the station to the point of intersection.

But if, as we recommend, stations be left on the main lines at stated intervals—say five chains apart, it is evident no difficulty will arise in determining the exact point of intersection, as well as the chainage, from one of those stations. The method of procedure by this plan is, to drive stakes at the determined distances—say at five or ten chains apart, and inscribe the chainage rudely thereon in Roman characters; thus, if stakes are left at every five chains, call five chains I., ten chains II., and so on; if the stakes are left at ten chains apart, call ten chains I., twenty chains II., and so on. Care must be taken in the measurements that the leader, in drawing the chain, plumbs back to the last mark at each chain's length, while the follower keeps the forward end of the chain well on with his object; and, as the marks employed in ranging out the lines are usually planted on hedge banks to render them at once conspicuous and less liable to removal, it will require very great care, in measuring past them, not to alter their position.

By attending to the foregoing directions, it is impossible that any difficulty can arise in the measurement of a line but what may be easily overcome. We will therefore suppose the first line of a triangle to be measured, and will proceed to the measurement of the other two sides. At any spot supposed a good termination for the first line, set up a mark, and, if sufficiently acquainted with the locality, a new line may be immediately commenced therefrom, either by taking a forward or backward mark in the determined direction. But, if the surveyor should not possess the requisite knowledge of the locality, it will be necessary for him to explore the direction in which he contemplates running his line, previous to setting it out, and in doing so it will be found advantageous, roughly to range out his track, but which will occasion little loss of time, as the marks employed, from being properly distributed over the ground, can be quickly placed in correct position when the direction is finally determined on. It will probably be the better way to commence ranging out the direction of this line from its expected termination, towards its commencement, not troubling about the point of intersection so long that it is sufficiently near to the extremity of the previous line, and free from any positive objection. The first thing to be attended to, when this point is decided, is, to determine the chainage at the intersection, and make a permanent station; the advantages (if any) of backing the line beyond the intersection may then be considered, and, if favourable, the correct backward direction may be noted, in case of the removal of the forward marks. When these things are done, the measurement of the second main line may be commenced, and carried through, in precisely the same manner as directed for the first. On completing the measurement of the second line, the same operation is to be pursued for determining the direction of the third as for that of the second; but some attention will be requisite in keeping the direction within the limit of the first line, to prevent a *minus* quantity being added thereto, as pointed out at page 44. Where the angle between the first and second line is taken with a good theodolite, the direction of the third line, to intersect the first at any required point, is easily arrived at by a simple calculation in trigonometry, and which, in some localities, is a matter of importance. As we have previously remarked that the lengths of lines intersecting obliquely are involved in doubt, we may here observe that, where two sides of a triangle intersect obliquely, the error or doubt arising therefrom may be cleared, by running a line from the vertex, so as to fall into the base nearly at right angles, then there will be three sides of a triangle to find the angle at the base, the supplement of which will be the angle of the base in the other triangle. The two sides—*i. e.*, the common side and the intersected portion of the base—and the included angle being then given, the subtense is easily found, which should correspond with the measurement, if both are correctly performed. When measuring these side lines, stakes should be driven at regular intervals, and numbered in the same manner as directed for the first line; and we would direct the reader's particular attention to the subject of running several secondary lines from one main line to another, before commencing to fill in the detail, and which will, in the end, economize time. When closing on any one of the lines staked out in the manner directed, measure up to the point of intersection from one of the stakes, and form a station, entering the measurement of both lines in the field-book.

The great point to be aimed at, in laying out the lines of a parish survey, is, to inscribe or circumscribe the whole area with as few lines as possible, and in keeping them within offset distance of the boundary; but as this is scarcely possible to do in all cases, small triangles will require to be occasionally thrown out to enclose excluded portions, similar to those marked T T T in the diagram at page 51; and lines forming secondary triangles will be required, to take up the boundary where it falls far within the main lines, similar to T in the lower part of the same diagram; and, in throwing out triangles to enclose any part of the parish without the side lines, it will be necessary, where the chain only is employed, to produce the sides of those triangles until they intersect some one of the internal lines. Previous to the filling in of any part of the survey, it will be prudent to obtain such portion of the boundary and detail as lay without the main lines on that part intended to be first filled in; and, when this is done, internal lines may be run for the detail;—the surveyor strictly confining himself to one portion only of the survey, and entirely filling it in before commencing on any other part,—his work will then never get confused. With regard to the direction in which these internal lines are to be run, no instructions can be given, except that an oblique *close* is at all times to be avoided.

We may here observe that, if the angles of the first few internal lines are taken, their position will be fixed without regard to their measurement, and, on being laid off on the plan, the distance at which they intersect other lines will correspond with the measurement in the field, if all is correct; but if, on protraction of the angle, it should not pass exactly through the point, as determined by measurement, it should be made to do so, as more dependence is to be placed on the distance than the angle in such a case; but, by taking the angle, any error committed in the register or in measuring from any one of the stakes will be immediately detected. And we may further remark that, if the surveyor is expert in the use of the sextant, it would be very desirable to have the angles of all the internal lines taken, except they are well tied; but where their position is determined by intersection of their extremities only, the angle is requisite to insure its correct position, especially if the intersection approaches a right angle. In every case where field work has to be transferred on to paper, it is necessary that the operation should be performed at as early a period as possible after the measurement: and the necessity for so doing is obvious, for, in the event of an error being committed in the field, the circumstances attending the measurement will be fresh in the surveyor's memory, and he will be enabled to rectify it with little trouble; but where time cannot be spared to transfer the offsets, &c., to paper, the lines of construction should certainly be laid down after each day's work.

With regard to the method to be pursued in laying down the main lines of a large survey, some consideration is required. It is generally performed thus:—The longest line is first drawn in, and its length carefully marked off with the scale. At the point where one side intersected the base, the edge of a broad and stout strip of drawing paper is fixed with a needle, and the length of the lines marked on it; a fine-pointed pencil is then applied to the extremity of the line, and the paper strip made to revolve from the needle point, describing an arc of a circle; and, in a similar manner, another arc of a circle, intersecting the first, is described from the other point; and to the intersection of the two arcs the lines are drawn. A much more correct method is, to mark the lengths of all the lines on the bevelled side of a thin straight-edge, lay off the base, and sweep in the sides as before; and the lines can be at the same time drawn in, before removing the straight-edge. Beam compasses are also commonly employed for the above purpose in small surveys; and as they may be had finely graduated, with a vernier to read fractional parts, very great accuracy may be arrived at; but, for large surveys, they are generally of little use, from their shortness of beam. We have, however, seen the beams made of several pieces of brass tubing, so as to reach to any required length with perfect steadiness. A beam of fir, of any required length, fitted to the points by a village carpenter, is also often used; but, in such case, the lengths of the lines first require to be laid down on some convenient spot, and then transferred to the paper. But when the point of intersection of the side lines is obtained, considerable difficulty is often experienced in drawing in the lines perfectly straight, from the want of a straight-edge of sufficient length to lie over the two intersections. When such is the case, and the paper free from *buckle*, a good substitute will be found in a thread blackened with a burnt cork, and snapped when correctly over the points. When the angles of the main lines are taken, we would advise, as a preferable method, that they be laid down by their sides, in the manner described, in place of by their angles; but the protraction might afterwards be advantageously employed in checking their position; but these few last remarks on angles more properly belong to another part of our treatise, to which the reader is referred. We have now gone fully into the subject of parish surveying; indeed, more so than the limits of our work will properly admit of, but, as the directions we have given on the several matters brought forward are strictly applicable to every description of land surveying, except that previously excepted, we have been induced to do so, seeing that many remarks which would have been necessary in our further investigation of the subject will not now be required.*

* The price of surveying work, Captain Dawson observes in his report, "varies considerably, according to the nature of the ground, and the quantity of detail to be represented. From numerous inquiries which I have made, it ranges generally from sixpence to a shilling per acre, though in some places it may be done for twopence, (!) and in very complicated districts would cost eighteen-pence. Ninepence per acre may be considered a fair average price for surveying work throughout the country, including a plan on the scale of three chains to an inch, and a book of reference. The expense of copying plans on the three-chain scale, in the best manner (exclusive of stationery), varies from a penny to threepence per acre. Twopence per acre may be considered a fair average price for a copy of the plan and book of reference; and the cost of copying a plan on the same as the original will not, in general, exceed the cost of reducing it to the scale of six or nine chains to an inch.

We now proceed to the matter of railway surveying, the practice of which differs considerably from ordinary surveying; indeed, so much so, that a very good and expert surveyor of estates or parishes will often make a very indifferent or slow surveyor for railways. From the survey of property required for such a work being long and narrow, the ordinary method of triangulation cannot be resorted to; and the grand object in the first instance being position and general direction, rather than scrupulous accuracy of content, a series of continuous base lines form the principal feature in the survey. It is desirable to obtain the relative direction of these base lines with the utmost nicety, and which is generally ascertained by means of a good theodolite, although it *is barely possible* that accuracy may be arrived at in the matter with the chain alone. When the positions of the base lines are determined on, ranged out, and the measurements completed, the enclosures through which the intended road is to pass will occupy attention: it is usual, and indeed necessary, where the chain alone is employed, to triangulate the enclosures, closing as often as possible on the base, and intersecting it with cross lines wherever it can be done, as, by such means, when a point on one side of the base is fixed, a line extended therefrom beyond the base on the other side, to any part where the chainage is noted also, becomes a fixed point. It is, however, exceedingly tedious to make a survey of this kind without using angular instruments, and, we may say, liable to great inaccuracies. The most useful instrument for such surveys—and, indeed, for every description of surveying,

The best drawing paper, mounted on good strong linen, costs from sixpence to eightpence per square foot. The expense of reducing, transferring to stone, and striking off fifty-eight copies of the map of a parish containing 4,733 acres, would, it appears, be less than eight shillings a copy."

is the pocket-sextant, of which we give a full description in our account of surveying instruments. Where an angular instrument is employed in this description of work, the enclosures are all surveyed by trapezium lines, except in some cases where the position is such as to invite the formation of a triangle, and which we would always recommend to be done where it can be accomplished without extra trouble.

The survey represented at plate 7 is for a railway, as will be perceived on inspection of the plan. The first operation performed was, to set out the base line A B C with short ranging poles of about five feet in length, which were planted successively in the several hedge rows, a breach or gap being made in each, where the line intersected it, to allow of a clear sight from end to end. The ranging out of such a line is generally performed with the theodolite, which is occasionally planted, and the axis of the telescope, or line of collimation (see description of the theodolite), brought exactly in the plane of the base line, and, by the instrument being set level, the required number of poles or other marks may be fixed in the ground in a perfect vertical plane, at least as near as mechanical aid can accomplish it. The base line being ranged out, the measurement of it was commenced from the station adjoining A, where a ranging pole with a small flag was originally set up. Before arriving at the first fence with the chaining, a false station, a, was put down, and the exact distance (540) entered in the field-book, and also the crossing of the fence a little beyond; further on, another false station, b, was entered (1330), and also the crossing of the fence near thereto; and in a similar manner were successively entered the other false stations, c (2240), d (3380), e (4080), and B (4720), as well as the several intermediate fences. It is the practice of many surveyors, after measuring out a certain length of base—as A B, for instance, to commence at B, and work back to A, by which means a saving of time is effected; but, as such method might perhaps perplex the beginner, we return to A, to commence operations on the detail. At this point, the first thing to be done is to lay out a side line, A f, and take the angle (70° 20′) f A a; but, in laying out this line, care must be taken so to choose the position of the station f, that the line f A can be backed or produced to g. When this is done, retire to g—which will of course be in a line with the marks set up at A and f,—and from thence *commence* the measurement towards f, taking the necessary offsets, and carefully noting the distance g A, intersecting the base line. When the measurement is carried to the top of the field, fix on a station, f, which must be in a position to allow of the longest and most favourable line being laid out for taking up the adjoining fences, as f h i j; when this point is settled, the measurement—if the chain does not reach so far—is to be continued beyond f, until it cuts the fence at 600. Then take the angle A f h (108° 10′), and measure the line f j; on arrival at about h (480), form a station in the most convenient position for passing a line down the adjoining fence, such as h a l k; and take the angle f h a (96°), or j h a (84°), either of which is immaterial. Now this line may be measured before proceeding further with f j, or it may be deferred until afterwards; but, in the present case, we pursue the former method. The line h k is therefore measured forthwith, and the chainage intersecting the base at a (240) noted, and the measurement carried forward to k, leaving an intermediate station, l, to be used for the line l m n, in taking up the adjoining fences. From the termination of the line under measurement at k, take a new line to g, which will be a close, and finish the measurement of the first field. It will be perceived that no positive necessity exists for taking the angle h k g, as, the two points k and g being fixed by the former angles and measurements, the distance k g will not come in if an error has been committed, except it intersects one of the lines at about a right angle, in which case the angle will be required to verify the measurement; it is also usual to take the angle of the cross line, off the base, as h a A, to check the angle observed at h. We return now to h, and continue the measurement onward towards j; when arrived at about i (1220), take the angle j i b (83° 50′), or h i b (96° 10′); but, before measuring this line, it will be best to complete the line under measurement up to j, where leave a mark at even and convenient chainage. Then return back to i, and measure a line i m, cutting the base at b (220), and continue it onward to m (615), where leave a temporary station; the angle at b, off the base line, should also be observed, which would determine the position of i, and check any little inaccuracies which might have occurred in the previous *angles or measurements.* Then measure a line l n, cutting the last line at m, and noting the intersection; from n—which will be put down in a line with the previous formed stations, c and j—measure a line, n o, and note the intersections at c and j, and at c take the angle off the base, which will fix the position, in plotting, of the lines f j, and n o; this last line is to be extended to any convenient length, so as to command a line, o p q, to take up the adjoining fences. Take the angle n o q (83° 18′), and measure the distance—leaving a station at p (1135)—to q (2535), where take the angle o q s (88° 15′). Measure q s, noting the chainage on closing at B, and also take the angle at the intersection; then carry on the measurement to s (1000), leaving an intermediate station at r (675) to measure a line from, as r e t, which measure, noting the close on the base at e—where the angle need not be taken from the line being fixed by three points—and form a station at t, in the same line with p and d. Then return to the station at n, and measure a line to e on the base; when at w, in a line with d t p, leave a station, w. From p measure a line to v, cutting the previous formed stations at t, d, and w; from v measure a line v s—which might be extended to u, and take the angle at s, from s B. This last measurement finishes the survey for the railway up to the road, which we think is sufficient to show the reader the method in which such operations are conducted. Where a person is familiar with this kind of surveying, much of the labour which we have detailed may be saved, as in many instances he might run his lines backwards or forwards, without walking to a distant part to resume his work; neither is it *absolutely necessary* that so many angles should be taken as we have mentioned, as those at h, i, and c, as well as some others, might have been omitted, for the direction of the line f j would be determined by the angle taken at f, and the direction of the line n o by the fixed points c and j. But the advantage of taking the additional angles is very great, insuring as it does the accuracy of the work, and, in the event of an error being committed, instantly showing where it has been made. Supposing that the false station at i was entered in the field-book twenty or thirty links in error, by taking the angle at that station it would be detected; but, if the angle was not taken, the measurement, i b, closing on the base, would scarcely show any difference from the correct distance; but, on producing the line to m, the error would be very serious, distorting the whole survey, but still remaining undetected. The station at i being moved forward twenty links to the black line, would scarcely affect the distance i b, and which would be assumed correct; the line i m would be consequently produced through b in the direction of ✱, and the station m set off on that line would be also considered correct; but it will be seen, on reference to the plan, how very incorrect the station at m would be placed. It might, however, be said that the measurement of m l would correct this; but suppose the error of twenty links to have been committed at h, and which would probably affect i in the same quantity, the station l would then be nearly as much in error as m, and the line m l would plot correctly as to length; the distance k g would, it is true, correct the position of k, but, in the general method of surveying, the distance k g would be necessary to fix the point g, and a continuous line, g A f, being frequently not adopted, the whole survey would in such case be twisted out of its proper position.

We have considered it useless to proceed further than the road with this system of surveying, which we think will be found quite sufficient as an example. The remainder of the plan, from B to C, is illustrative of the method pursued in effecting such surveys with the chain alone; but, previous to our giving any description relative thereto, we have a few remarks to make on that part of the survey which we have just described. The field-book is on the sketch system, as explained in a previous example. The angular instrument employed was the pocket-sextant,—which, of all the surveying instruments, is the most useful,—the theodolite not being required, from the open and level character of the country; but if the country had been unlevel, and thickly wooded, it would have been essential in setting out the base line; but of this matter we shall have occasion to treat, under the head "of observing with the theodolite," to which part of our work the reader is referred, as well as the method of laying down the lines of construction by the angles.

In carrying out a survey for a railway—as exhibited in plate 7—we have before remarked that very many more lines are required than when an angular instrument is employed; and, as we have laid down the first part of the above plan by instrumental construction, we will complete it, as an example, by the chain method. We have already described the construction of the plan with the sextant up to the road at B, from which we commence the description of the remaining part by the chain method, as though it was the commencement of a survey, and the part A B unconnected with it. In the first instance, the base line, B C, would be ranged out and measured in the usual manner, and also a line s B y; but to fix the position of this last line, with regard to the base, it becomes necessary to measure y x, to form a triangle B y x, otherwise it could not be plotted. A line, x s, is also required to be measured—forming a triangle x s B, to check the position of the line, y s, previously determined by the intersection at y. These two lines, it will be perceived, are wholly useless in so far as regards the actual delineating of the features of the survey, but are still necessary, to fix the two points, y and s, to throw out side lines from—as s u and y z *a b*—to take up the bounding fences. When the line x s is closed, measure s u, and from u measure u z, cutting the base at x; at z leave a station, and go back to y, and measure a line, y z *a b*. From *b* run a line to *i*, leaving a station at *h*, and intersecting the base at *g*; from *i* measure a line to *f*, and from *f* to the station left at *a*; leaving stations at *e* and *c*, and intersecting the base at *d*. Then go back to z, and measure z *c* and *c g*; from *g* measure a line to the station at *u*, intersecting *e;* return then to *i*, and run a line across the base at *k* to about *n*, where leave a station; from *p* on the base, measure *p o*, cutting the station *n;* also measure *o b*, leaving a station at *l*. Also measure *h l*, *l k m*, and *i m p*, and *k h*, which completes the survey beyond the chase; and, as the continuation would be little more than a repetition of what we have just given, we think it unnecessary to proceed further with the description. It will not fail to be perceived that many more lines require to be measured by this plan than when an angular instrument is used—such as s x, and y x, *h k*, *k n*, and many others; but the reader will perceive, on looking over the remainder of the plan, which are the extra lines; and, although the survey might be plotted without several of these extra lines, they are all necessary either as principals or checks. The plotting of such a survey as that just described is performed by intersection, in the same manner as the single field at page 10.*

A common practice with many chain surveyors, to facilitate their work when employed on railways, is, to form triangles at the commencement and at intervals on their base line, and do little better than *fudge* in the intermediate parts,—or, at least, pay very little attention to the intermediate lines of construction—knowing that the errors which accumulate as their work advances will be stopped at the close on each successive triangle; a practice which cannot be too much discountenanced.

Of the advantages to be derived from the use of an-

* The price which has been usually paid for railway surveys, when contracted for, is from twelve to fifteen pounds per mile, including plans and books of reference, which is equal to from eighteen-pence to about twenty-two-pence per acre. Surveyors' remuneration by time has ranged from one to three guineas per day, exclusive of expenses, according to the emergency of the work, and expertness of the individual.

gular instruments in performing surveys for railways, &c., we have only to observe, that correct surveys may be made therewith during the summer months; which, under ordinary circumstances, is never attempted, in consequence of the great injury inflicted on farmers by damaging their corn crops. The principal part of the damage is done when measuring the base line; and as such a line is absolutely necessary in accomplishing any extensive survey, the only method which suggests itself is, to set out and make use of a base line, without subjecting it to direct measurement. The manner in which this may be performed is, by first ranging out a base line for the total distance, by means of marks planted in the hedge-rows, and in fields not corn cropped; circumscribing the enclosures with as few lines as possible, and taking the subtended angles, and also the angle *off* the base, of the several lines which intersect it. Such portions of the base are to be measured where it can be done without injury; the intermediate portions can then be easily deduced by calculation, or obtained by plotting.

We have known several extensive and correct surveys performed in this manner, which, by the usual method, it would have been impossible to accomplish.

Straight fences running transverse to the base may be taken up from the *side* lines—without the aid of cross lines closing on the base—by what is termed "sighting," which is offsetting to the ends of the fences in the line of their direction; but, to be certain of accuracy, the angle should be taken at either end, which, on protraction, should both overlie the fence. This plan is much practised on parish surveys, and is both expeditious and accurate when judiciously employed.

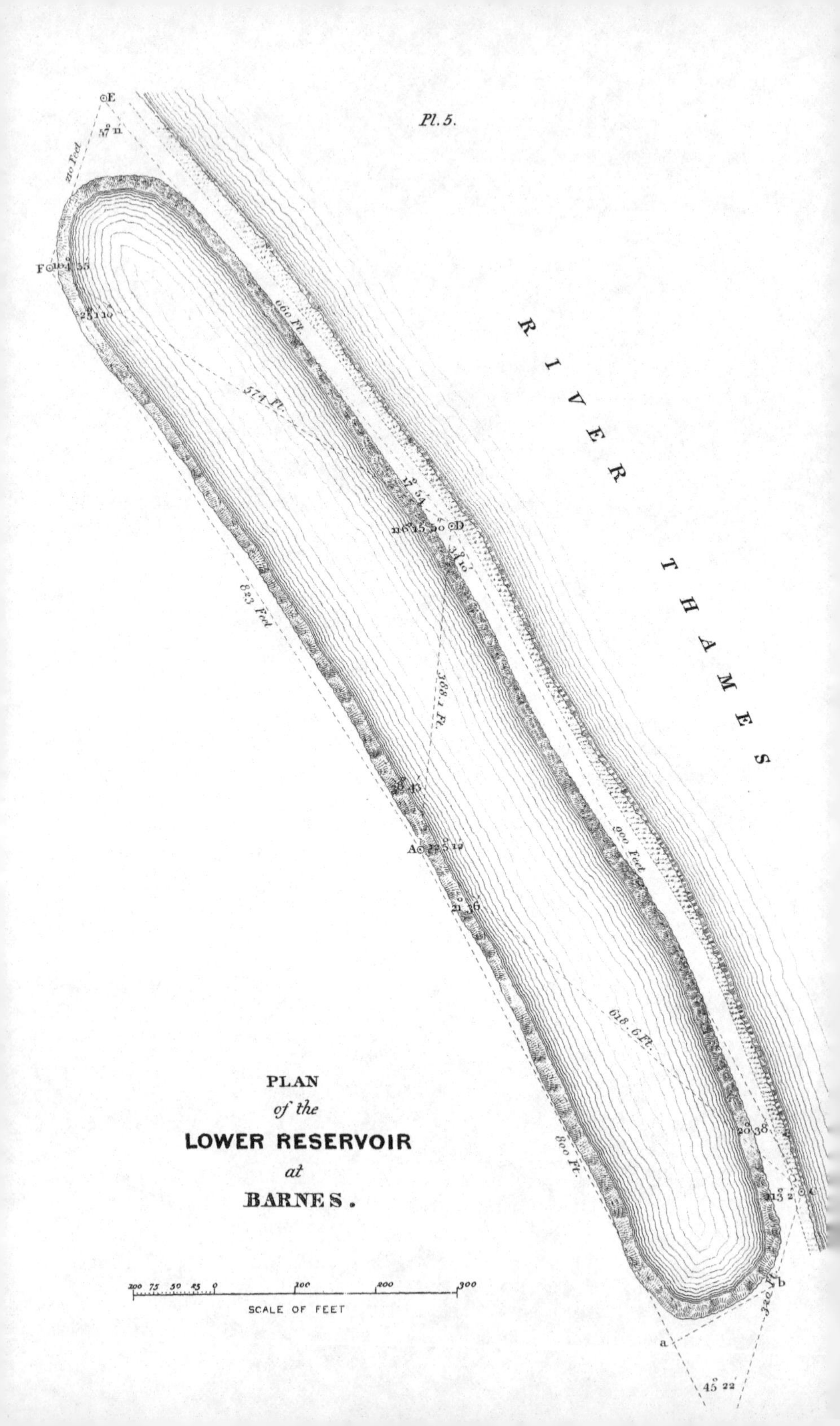

Pl. 5.
RIVER THAMES
PLAN
of the
LOWER RESERVOIR
at
BARNES.
SCALE OF FEET

CHAPTER V.

SURVEYING WITH ANGULAR INSTRUMENTS.—SURVEY OF A RESERVOIR.—EFFICIENCY OF THE THEODOLITE IN SURVEYING.—ON MEASURING THE BACK ANGLE.—TOWN SURVEYING, AND SUGGESTED IMPROVEMENTS.—SUBTERRANEAN SURVEYING.—EXPLANATION OF A "TRAVERSE;" THE LATITUDE AND DEPARTURE.—HARBOUR AND COAST SURVEYING.—SOUNDINGS.—MEASURING INACCESSIBLE DISTANCES.—CONCLUDING REMARKS.

At the conclusion of the last chapter we gave a simple case of instrumental surveying, but in the present we intend going more fully into the subject; and shall illustrate our remarks by some examples which could not have been accurately performed without the aid of angular instruments. We say accurately, because a skilful surveyor can determine angles, inaccessible distances, and every other problem usually performed with angular instruments, by the chain alone; but the results so obtained are necessarily inaccurate, from the numerous additional lines required to be measured for such purposes; and, as angles can be more correctly measured with an instrument than the length of lines with the chain, it follows that the method by the chain must be proportionably incorrect.

The plan of a reservoir, represented in plate 5, forms part of a survey which we made some time since, in the parish of Barnes, in Surrey; and when we inform the reader that the river Thames flows in the same circuitous line, and within about forty feet of every part of its upper boundary—which obliged us to adopt so many lines—we think the necessity of a good angular instrument will be apparent, to have enabled us successfully to accomplish it. It was not only necessary in this survey that the general features should be strictly obtained, but that the lengths and breadths should be ascertained at every point with scrupulous exactness; and, as the reservoir was nearly filled with water, it was impossible to arrive at the desired results by any other means than that adopted, *viz.*, circumscribing it with as few lines as possible, and observing the subtended angles, at the same time checking their position by trigonometrical observations and calculations.

A point, A, was chosen, on one side of the reservoir, to commence from, where a theodolite was set up, and a line, A B, "boned out," and immediately measured; but, before the removal of the theodolite, the point B at 800 feet was carefully determined by means of a stump or picket, and also the point a, at 660 feet. A plumb line, suspended from the vertical axis of the theodolite, also determined the exact position of a picket to be driven at A, where, also, a short flag pole was erected. The instrument was then removed to B, and set up plumb over the previously driven picket, and a line, B C, "boned out" and measured as before; and at 320 feet another picket was placed in correct position by the theodolite, and carefully driven; on this line also, at 200 feet, a picket was driven, and a cross line, b a, measured, checking the angle to be taken at B; after which the angle A B C was observed, repeated, and a mean registered. The theodolite was then taken to C, carefully set up plumb over the picket, and the line C D "boned out" and afterwards measured; the point D at 900 feet being determined at the same time by a carefully driven picket, guided by the theodolite as before. From C, the angles B C A and B C D were observed, repeated, and a mean registered as at B. Also, in a similar manner, was the instrument planted at D, and the angles C D A, A D F, and F D E taken; at E, D E A; at F, E F D and D F A; the last angle being taken to the point of commencement, which, of course, on being laid off on the plan, would pass through that point, if correct, in addition to the measurement of 823 feet exactly closing. At A the observed angles were F A D, D A C, and C A B, which closed the operations. It will be perceived that the distance across the reservoir, from one station to another, may be calculated from a great variety of data, which, in the event of any error in the angles or lines, could have been readily corrected, without having recourse to further operations in the field. A check was also obtained on the instrumental part of the survey; for, as all the *internal* angles of a rectilineal polygon, together with four right angles, are equal to twice as many right angles as the figure has sides, it was only necessary to cast up the observed angles; and, on the number of sides being multiplied by 180°, and 360° deducted therefrom, both results would be the same, if correct. Another reservoir was included in our survey, but which we have here omitted, from the similarity of operations in both cases. We have not thought it necessary to give the detailed field-book, but, on reference to the plotted plan, there will be found the lengths of the measured lines, and also the observed angles, from which data the distances across it were calculated.

The use of the theodolite, in the common practice of surveying, is confined chiefly to obtaining the subtended angles of the principal lines of construction, after which it is generally laid on one side, and a more portable instrument employed for the minor lines. Frequently, in these subsequent operations, no other instrument than the chain is used, in consequence of the cumbrousness of the instrument, which makes an extra assistant necessary, and also from the time required at each station from whence angles are taken, in setting it up plumb over the angular point, levelling it, bisecting the observed points, and reading off the angle. But it is our opinion that the theodolite might be extensively and beneficially employed, in parish and other large surveys, in fixing numerous advantageous points trigonometrically from a carefully measured base; and from any one of those points, so determined, lines could be measured for any short distance in the direction of any determined point, for the purpose of filling in the detail, without the necessity of closing on any other line; and, for the purpose of verification, an angle might be taken from the termination of such line to some other previously determined point. In railway, canal, and other similar surveys, we have before remarked, that the theodolite is used to range out and take the junction angles of the base lines; the detail of the work, *i. e.*, the delineation of the fences, &c., being performed with the chain and sextant, or the chain alone. In laying out the base lines, distant permanent objects should always be obtained, if possible; and when such is the case, and it becomes necessary to lay out a new base before arriving at the object with which the line is on, it will be advisable to take the angle,—not from off the line *backwards*, but from the forward object;—and, as a check on the angle, the previous line might be continued onward a few chains, and connected by measurement with a station on the new line at the same distance from the angular point, and which, in plotting, will test the accuracy of the angle, and subsequent operations connected therewith. If it is considered necessary, the angle of subtension may be directly deduced from these measurements, forming, as they do, an isosceles triangle; for, as either of the equal sides is to half the subtended distance, so is radius to sine of *half* the angle of subtension; and the whole angle, subtracted from 180°, will give the forward angle, which will be its supplement. While on the subject of measuring the subtended angles of base lines, we may mention a practice which we invariably pursue in an undulating country, and which we are satisfied is of more consequence than at the first sight may appear to the reader. It is, to commence and terminate such lines on high or rising ground, in every instance where practicable, for the purpose of being better able to obtain a good *bearing* on the back line, when taking the angle of the forward base. The case under notice is a road survey, requiring a long narrow strip of country to be delineated, in which, as we have before observed, there is no opportunity—at least, without great trouble and loss of time—of checking the relative position of the base lines by triangulation; and from this it will be evident, that the absolute relative positions of the two lines, and, consequently, the value of the survey, will depend upon the extreme accuracy with which the subtended angle is measured. No doubt can be entertained but that great advantages are gained in surveying, if the one extremity of the base line can be seen from the other, as in such cases little harm would arise from a slight deviation from a straight line in the intermediate part; but if a very slight intermediate deviation, indeed, from a straight line should take place when the one extremity cannot be seen from the other, the angle would probably have to be taken from the deflected part; if so, it will immediately be seen how very erroneous the relative position of the two lines would be placed, although the angle be measured with ever such care. In the example of which we are about to speak, it will be seen that the error arising from being able to obtain only a very short bearing on the back line can, in most cases, be avoided, and the angle taken with as much accuracy as if the extremity of that line was visible from the angular point, and with this additional advantage, that a check (not very severe it must be admitted) is obtained by triangulation. In the accompanying diagram, the base

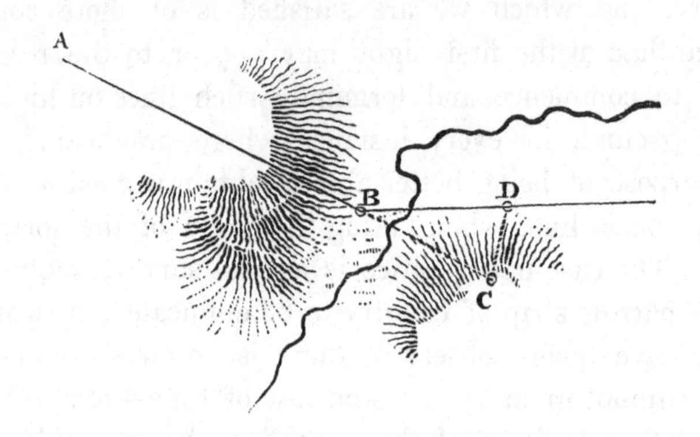

line A B is required to terminate at the point B, to allow the line B D to be laid out advantageously. It must be understood that the station at B is on low ground, with the crest of a hill rising immediately behind it, by which the sight from B towards A is limited to a very short distance. and consequently the back angle, if taken from the point B, will have but little *bearing* on the line B A, in which case, although the operator be ever so expert in the use of angular instruments, or careful in recording his observation, no very great dependence could be placed on the angle A B D, which it was the object to determine. The method proposed for the purpose of accurately obtaining the subtended angle A B D, is to extend the line (trace it only, not measure it) from B onwards to a point C on high ground, or until the point A, or other well determined point, on the back line become visible.

At C set up a theodolite, and by slightly removing the same to the right or left the correct extension of the line from A, through some well determined points, to C can be easily determined. This being done, the station at B can be precisely fixed with the instrument from C, (a plummet, suspended from the axis of the instrument, will also as precisely fix the point C), which being done, an angle may be taken to any object on the new line, as to D. The instrument being then removed from C (a flag being fixed up in its place) to the before determined angular spot B, the angle D B C will be taken, which is the supplement of the required angle A B D. When measuring the line B D, if the distance B D be noted, and the angle B C D taken, a check on the accuracy of the work will be obtained, for the three angles will, of course, amount to 180° (the sphericity of the globe in such operations will be inappreciable), and having the length of one side, of course the other sides, and angles, may be obtained by calculation. By this means the angle between two base lines can be measured with nearly as much precision when the back angular point is invisible, as on the contrary.

In effecting some railway surveys, where, from the difficult features of the country, or contemplated opposition to its construction, a greater width than usual was required,* we have known a series of two, and sometimes three, base lines laid out parallel with each other, or nearly so, for the total distance requiring such width; the angles of which, in continuation, were determined in the same manner as when a single base line was em-

* The usual width, in an ordinary level enclosed country, is a quarter of a mile for the detail plan, but we have sometimes known more than half a mile in width to be included in such survey. In hilly districts, or where there is much ornamental property, buildings, &c., the position of the line will of course be more strictly defined, and less width be required.

ployed; but as they deflected at points nearly abreast, a tye line was measured across, connecting the angular points on each line; and from this tye line the angle of each forward base was taken, in addition to the back angle of each previous connected base, by which means they were not only laid out correctly in continuation, but their parallelism was efficiently checked. In filling in the detail, where such a system is pursued, nothing but the chain will be required, except in very rare cases indeed; for the principal cross lines closing on three points, cannot fail in being correct, and may with safety be produced beyond the outer base lines to take up any required features. An advantage also presents itself in this system of surveying, which is important; it is the great facility and undoubted accuracy with which external objects, however distant, can be laid down on the plan without additional measurement; thus, in the accompanying diagram, suppose marks to be left at f and e when

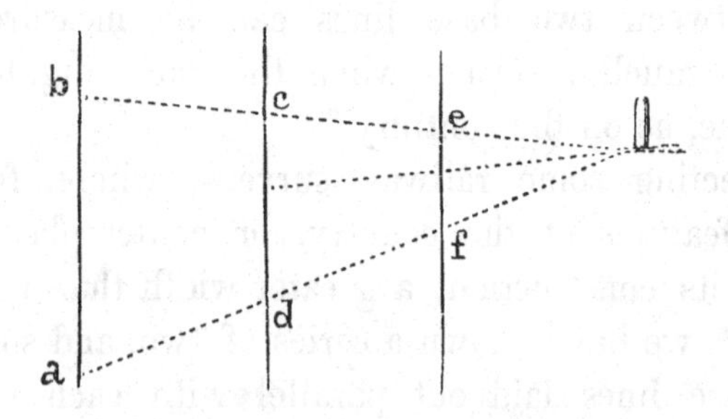

measuring the first line, and marks d and c on the next, *on* with the previous marks and a tower, the position of which was required; it is evident that the desired object would be obtained by drawing lines on the plan from the marks at d and c, through e and f, until they intersected, which point would correctly represent the situation of the tower; but as in this case there would be no proof of its accuracy, it would be necessary to notice the points at a and b, when measuring the third line; or, if two base lines only were made use of in the survey, a third point on each line would prove the previous intersection.

In taking plans of towns and large villages, the theodolite is particularly valuable; and, indeed, without it, such surveys could not be made at all,—at least, with any pretensions to accuracy. In the case of a public work to be constructed within the precincts of a town—such as the Saint Katharine's Docks, or of a work that penetrates far into a town—as the Greenwich, Blackwall, or Eastern Counties' Railway, the approaches to the New London Bridge, &c., a survey of the district through which it passes is required; and which must be of the most minute character and accuracy,—correctly detailing every street, court, alley, dwelling-house, yard, and appurtenances, on the contemplated line of operations, as well as those on each side, which may possibly be affected thereby. Any system of triangulation, in such cases, is evidently out of the question; and the only method of procedure is, to form a network, by measured lines passing through two or three parallel lines of street in the required direction, if so many can be made available; and connected, wherever an opportunity offers, by cross lines along the streets running transversely to the others. The direction of all these lines, the one from the other, is carefully obtained, while the measurement is proceeding, by means of the theodolite, planted at each intersection. Where a series of quadrilateral figures can thus be connected together, or rather built, the one upon the other, throughout the survey, little chance of error can arise but what will be detected in plotting, and, consequently, may be easily corrected; but it will happen very frequently, that only a single line can be carried forward in the required direction which will connect with the previous operations, without incurring immense labour in carrying lines very far beyond the bounds of the required survey. In such case, without the labour of the additional lines are incurred, the *whole* dependence will rest on one observed angle, and which, we need scarcely remark, would require the greatest care in the setting up of the instrument, bisecting the object, and in noting and repeating the angle.* Some slight check would, however, be obtained in the correct direction of the lines, by noting the bearing by the compass; sufficient, indeed, to prevent any gross errors being committed, but not sufficient to insure accuracy. But, as the purchasing and pulling down property on the line of improvement, as well as erecting the required works, are necessarily carried on in detached portions, of course the only method of determining the points of operation is, by reference to the plan, on which the requisite particulars will be drawn. Now, in the case of a railway viaduct penetrating far into a town, the property is purchased from the plan, at the points where the works are first required to be commenced—perhaps at twenty different places—all of which may be in a right line, according to the plan; but, in execution, when the intermediate property is cleared away, it may happen that no point between the extremes, where buildings may have been razed, and works erected, is correct, thereby occasioning serious alterations,

* In making a town survey, we have sometimes measured lines without the limit of our plan, exceeding in the aggregate a mile and a half in length, besides taking three or four angles in that distance, for the purpose of closing on a particular point in the survey, only determined by one previous angle; and have been frequently as much surprised as pleased, on joining, to find the angle and distance exactly intersect it.

or total reconstruction; an occurrence by no means uncommon.

From what we have just observed, the necessity of extreme carefulness in making such surveys will be apparent; and, therefore, in the practical details which we are about to give, we shall not reiterate the injunction. To make a good survey of a town, or of part of a town, the surveyor should possess a good theodolite, of not less than five inches in diameter; and have at least four intelligent persons as assistants; half a dozen light rods, of about six feet in length, with small white flags fixed on one end, and the other end shod with iron; a one-hundred feet tape, and an annealed chain of similar length, also divided into feet, with a few spike nails to drive into the ground at the angular points. Being thus provided, the surveyor is to proceed along the line of operations, and fix on the point of commencement, not at one extremity, as in common surveys of open country, but at one end of the longest street running in the longitudinal direction of the survey, and which may be favourably situated in other respects. If it lies about midway between the extremes of the survey, it will be the more desirable; for, in surveys of this kind, the most favourable figure which can be constructed is a trapezium; and as an infinite variety of trapeziums can be constructed with the same sides, it follows, that the angles are indispensable in determining their position. Now, as in direct lineal measurements, so in the observing of angles, we only arrive at an approximately correct result; and as in this kind of surveying we keep adding trapezium on to trapezium—in each addition assuming the previous constructed figure to be correctly posited, it follows, that these inappreciable errors must keep multiplying, and tending to throw each additional figure more and more from its correct position; and the greater the number, or the greater the distance over which the operations extend, so will the errors of position increase. But if the work is commenced from a base line situated midway between the extremes, and supposing an error, however slight—and we have endeavoured to show that such operations are invariably attended with error—to be committed at the commencement, its effects will extend only over one half of the survey, and the other half will be unaffected thereby. The same beneficial results may also be obtained by *commencing* to plot from about midway of the survey, towards each extremity, which in many cases will be found a preferable method to the former, which involves the necessity of reference to the original stations and lines on commencing the second portion of survey. This matter being decided on, the surveyor is to set up his theodolite at a junction of streets, and bone out a line along each,—until terminated by some physical obstacle, —and take the subtended angle; and, in general, it will be found a good plan, to mark the position of the line by a vertical chalk mark on the object obstructing the further extension of it, or otherwise, by taking some well-defined mark—such as the door-post, window-frame, or some angular point of a building, &c. Along this line the assistants are to fix a few intermediate flag poles, to facilitate the direct measurement, which is to be immediately proceeded with, after marking the position occupied by the vertical axis of the instrument by a spike nail or other means, and setting up a flag pole to take the back angles to, when arrived at the termination of the respective lines. In measuring these lines, offsets are to be taken to all bends occurring on either side of the streets, and the chainage, when opposite to the division or party wall of each house, entered in the field-book, but it will not be requisite to take offsets to such divisions, except where the face of a house deflects from a right line at such point. Offsets are of course requisite to the angles or corners of all side streets, alleys, courts, &c., but as such measurements, although intended, are not always made at right angles, it will be necessary to measure the *widths* of such side streets, &c., in case the two offsets determining the distance from the chain line should not have been measured *parallel* to each other. Stations are to be formed when opposite any side street requiring to be included in the survey; and, after the completion of the line under measurement, are to be referred to in succession—and lines measured therefrom, in the same manner as just described—the angle subtended with the line *off* which it is measured, being taken with the theodolite either to the forward or backward mark, or, better still, to both; which may, perhaps, prevent inaccuracies by the one angle checking the other, as both equal 180°. In this manner lines are to be laid out, one from another, until they *close*, generally forming trapeziums; and which are generally subdivided, by lines measured along the cross streets, into smaller trapeziums, checking the accuracy of the first great figure. In every case, we recommend the angle of each line to be taken; *i. e.*, in a four-sided figure, four angles to be taken; in a five-sided figure, five angles; and so on; and not be satisfied—as we have frequently seen the case—with taking two adjoining angles in a four-sided figure, and proving by the close of the fourth line, as in such case no positive check will be obtained. But if three angles are taken, a check will be obtained, as the angular line, when protracted, will pass through the fourth angular point, and the measurement exactly close thereon, if all is correct: but still it will not be so satisfactory as taking all four angles, or as many as the figure has sides. But if time is an object, and three angles considered sufficient, they should not be taken continuously from the commencement to the third angle, but one from each line centering in the first angular point; for when the third *continuous* angle is observed, the accumulated error at the second angle is still further increased at the third, where it will generally be found a sensible quantity; but by taking one angle on each side of the common angle, the error due to one angle only will be brought forward instead of two, thus avoiding the error attendant on one observation.

A great saving of time and expense might be effected in town surveying, as well as avoiding the disagreeable task of chaining a line in dirty weather along the streets, by using a good perambulator in place of the chain; and if a theodolite was fixed over the axis of the instrument, for observing the necessary angles, we think a most perfect machine would be the result. Certainly a perambulator of much better construction than those in present use would be necessary, but we conceive no difficulty would arise in making such an instrument to measure more accurately than the chain when commonly applied. An allowance would have to be made in measuring along a paved street, but the necessary correction might soon be arrived at by a few trials; and, where a line was measured on sloping ground, it would be an easy matter to take the vertical angle, and make the proper deductions from the measurement as determined by the perambulator.

Where a skeleton plan of a town only is required, *i. e.*, the streets, without the detail of houses, it may be very rapidly performed, by measuring two adjoining sides of a trapezium, and observing three angles; the length and position of the remaining two sides may then be determined, either by plotting or calculation, and the *fourth* angular point may be used, both in the field and office, as if determined by direct admeasurement. One half the time usually employed on such surveys would, by this means, be saved; and if the two sides were measured with a little extra care, a more correct plan would be produced. The only objection to the general adoption of this method is, in the taking up of side streets and deflections in the houses on each side the lines not subjected to direct admeasurement, and which—if such particulars were necessary—would require a separate operation.

Some authors and practitioners have recommended magnetic instruments to be used in making a town survey; but, from the diurnal variation and difficulty of guarding against local attraction, to say nothing of the trouble and uncertainty of checking the vibrations of the needle, and reading the magnetic angle when free therefrom,—we would not advise its use for such purposes, especially as the results given by the *limb* of the theodolite may be checked by the magnetic needle attached thereto, as fully explained in the description which we have given of that instrument.

Subterranean surveying, as practised in coal pits, mines, &c., is performed with a large compass, called a circumferenter, but more frequently with a modified instrument—which is half theodolite, half circumferenter, having the graduated limb and vernier of the former, with the large compass of the latter; by which arrangement the bearing or meridian angle can be obtained with much greater accuracy than with the common circumferenter alone. The method of surveying the workings or levels is "irregular," or by a traverse,—the position of the lines either entirely depending on the back angle, or on the angle subtended with the magnetic meridian, although it is a common practice to check the results given by the needle, by calculating the latitude and departure of the traverse, when, if the survey has been correctly performed, the sum of all the northings will be equal to the sum of all the southings, and the sum of all the eastings to the sum of all the westings. The accompanying diagram will fully explain the subject:—N S is a meridian line, and a *a* the latitude; and *a* b the departure of the traverse a b. Now it will be evident, on inspecting the diagram, that the sum of all the southings, a *a*+b S, is equal to the sum of all the northings, *b* d+d *c*+*d* a; and also is the sum of all the eastings. *a* b+*c* *d*, equal to the sum of all the westings, S c+c *b*. In the above figure, the position of the lines a b, b c, c d, &c., is determined by observing the angle each line subtends with the magnetic meridian, and measuring the length of the lines with a common chain; the latitude and departure of each station is then easily obtained by means of a traverse table, or by calculation; for there are given the hypothenuse, and the three angles of a right-angled triangle, to find the other two sides, *viz.*, the length a b given, and also the angle at a (and, from the angle at *a*

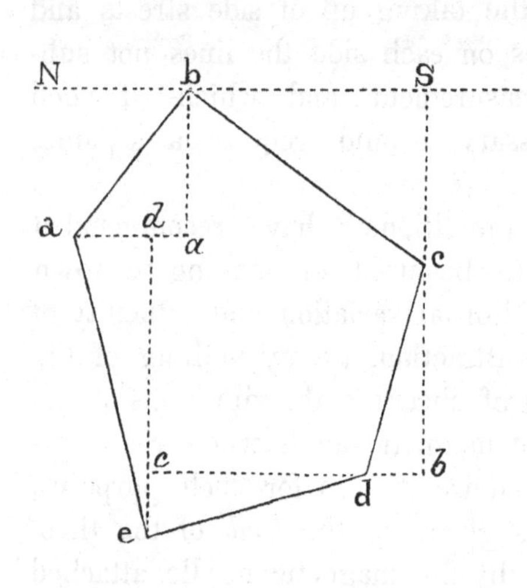

being a right angle, the angle a b *a* is known) to find the sides a *a* and *a* b. By bringing together the northings and the southings, and also the eastings and the westings, and subtracting their sums the one from the other, the latitude and departure of any station or point from another—such as e from a, supposing the line e a could not be measured—will be obtained, when the bearing and distance may be easily calculated; for a *d* the latitude, and *d* e the departure of e a, are two sides of the right-angled triangle, e a *d*; consequently the angle or bearing at e may be found thus:—as e d is to *d* a, so is radius to tang. ∠ e; and, to find the distance, say, as radius is to sec. ∠ e, so is e *d* to e a; or e a may be found by the well-known rule,

$$\sqrt{\overline{\mathrm{e}\,d}^2 + \overline{d\,\mathrm{a}}^2} = \mathrm{e\,a},$$

or, in other words, the square root of the sum of the squares of the two sides, e *d* and *d* a, is equal to the side a e, subtending the right angle. Plans of subterraneous surveys are also more correctly laid down on paper by the latitude and departure, than by the bearing of the traverse, and with much greater expedition. The bearing and distance of two stations remote from each other, may also be obtained by construction with a protractor and proper scale —first plotting all the lines by means of their angles and measured distances, and measuring the angle and distance of a line drawn on the plan from the one station to the other, with the same protractor and scale used in plotting. And, as surveys of pits or mines are generally made for the purpose of ascertaining the extent of workings in regard to adjoining property, for valuation of minerals, or for sinking shafts either for ventilation or convenience in raising the produce, it becomes necessary, subsequently, to trace on the surface of the ground, the lines measured beneath, when of course the reduction of several bearings and distances, into a single bearing and distance, to determine a particular point, is of great service, as, the fewer the operations to be performed, the less liability to error. We will now describe the method pursued in measuring the angles and distances of the several lines below ground, and also the manner in which they are subsequently traced on the surface. The instrument usually employed in taking angles, as we have before observed, is the circumferenter, which is planted plumb over the commencement and termination of every line, and the *forward* bearing taken, the distance being at the same time measured with the common chain. The method of procedure, in measuring these angles and lines, is thus:—At the bottom of the shaft or other determined spot, the circumferenter is set up, and a bearing taken in the required direction, to a lighted candle placed as far off as it can be seen. The distance is then measured in the following manner:—the circumferenter is removed, and one end of the chain held exactly over the angular point, or on the axis of the instrument. An assistant is then sent forward with the other end of the chain, holding a lighted candle in the same hand as the chain, and is directed by the leader, right or left, until the lighted candle (and end of the chain) is in a direct line with the light previously set up, the bearing of which was taken. The first chain's length is then marked, to which the leader advances, when the next chain is measured in the same manner as the first, and so on for as great a distance as required, correctly noting all the *bords*, water-courses, and other particulars. At the extremity of the first line the circumferenter is set up, and another bearing taken to an advanced light, and the distance measured as before, this process being repeated throughout the survey.

In tracing such a survey on the surface, showing all the windings of the several levels below, it is usual to go through the same processes of taking bearings and measuring distances in the required direction, except the whole are reduced to one bearing and distance, as previously explained, and which can then of course be set off in one operation; but in both cases there is generally some difficulty in commencing the operation above. Suppose the survey to have been commenced below, from the bottom of the shaft; in such case, to trace the workings on the surface, the circumferenter is planted as near to the shaft as can be done conveniently, and on the fore sight or north end of the instrument being set to the first bearing taken below, the back sight will cut the centre of the shaft, if the instrument is correctly placed, *i. e.*, if it is *on* the vertical plane of the line first run below; but when this is not the case, it is to be moved to the right or left, until it is in such position, when the back sight will intersect the pit, and the forward sight point out the direction in which the line is to be measured. At the extremity of this line—the measurement of which is to be commenced from the centre of the shaft—the circumferenter is again planted, and set to the next bearing, and the distance measured out as before; and so on for all the bearings and distances taken below. Another, and probably preferable method, is, to choose *any* spot near to the shaft as the point of commencement, from which all the bearings and distances are to be laid off in the exact order in which they were taken below; but before removing the circumferenter from the *first* station, the bearing and distance to the centre of the shaft is to be noted, which, being laid off from any point on the traverse, the extremity will be over the corresponding point in the pit.

A different method of using the circumferenter from that which we have just described is sometimes adopted; it is, to plant the instrument at alternate angles, and take back and fore sights: thus, in place of setting up the instrument at the commencement of the survey, and observing the forward bearing, it is set up at the extremity of the first line, and the bearing observed, but in a reverse order; or, in other words, the surveyor's eye is applied to the opposite sight-vane to that when a forward bearing is taken, the angle being read off from the south end of the needle instead of the north, as when noting a forward bearing. When plotting, these angles are easily reversed, and laid off as if taken from the north end, thus:—S. 10° 30′ West, on reversing, becomes N. 10° 30′ East. By this method the instrument has to be set up only half the number of times as by the former method, the forward bearings being noted at the same time as the backward angle. As these surveys form records which are being continually referred to from time to time, and additions made thereto as the workings are extended, it is necessary that all the bearings should have reference to the true meridian, in consequence of the *continued* variation of the magnetic meridian, and from two needles rarely having the same magnetic variation. The most correct method to determine the variation or declination of the needle is by astronomical observation made with proper instruments; but a good theodolite or circumferenter may be very successfully employed for such purpose, by observing the azimuth of the sun, or of a star at equal altitudes on different sides of the meridian—the mean reading will give the true meridian; equal altitudes being arrived at, at equal hours from noon, as 7 & 8 A.M., and 4 & 5 P.M. An approximate meridian line may be determined by setting up a rod in a vertical position, and observing its shadow at twelve o'clock; but it may be more correctly arrived at by describing an arc of a circle —with the length as radius—from the bottom of the rod, and noting the length of its shadow a few hours before noon. About the same time in the afternoon the shadow will again intersect the arc; the middle point, and the bottom of the rod, will then be a meridian line. Sometimes a drawing board is employed in laying down a meridian line; it is set up level, with a *stile* erected on one edge, several arcs of circles being described therefrom; at equal times, before and after noon, the shadow of the stile on these concentric arcs is marked, the central points on each being then on the true meridian. When a meridian line has been determined by these or any other methods, it should be permanently marked in the neighbourhood of the mines, for the purpose of determining the variation of any instrument which may be used in surveying the works. The present variation of the needle at London is 24° 6′ westward of north.* The diurnal variation has been observed to amount nearly to half a degree, but the usual variation is somewhat less than fifteen minutes.

To prevent attraction of the needle, when observing with the circumferenter, great care should be taken not to permit any iron to be within several feet of the instrument; but oxide of iron, or iron ore highly charged with sulphur, does not appear to affect the needle in any sensible quantity; neither do the iron tram plates generally

* In the former edition it was stated, that the variation of the needle was about 27° westward of north. The cause of the mistake was simply this:— the author was confined to his bed by sickness, and, to employ his mind, wrote the work as an amusement; but not remembering the exact variation, he employed a friend to inquire, who informed him that Captain Mudge had just completed some experiments, and decided the variation to be 27° 20′ west. The diurnal variation was, by mistake, printed a degree and a half in the former edition, instead of half a degree.

laid down in pits, when the instrument is elevated a few feet above them. In taking bearings on the surface, the instrument should never be set up near to any perpendicular objects, as trees, houses, and the like; as it has been discovered that all vertical objects have a north pole at the bottom and a south pole at the top, and that they exert a very sensible influence over the needle.

For an account of the method of observing with the circumferenter, and several problems connected with it, see description of the instrument.

On the subject of harbour and coast surveying we cannot be expected to speak very fully, as a full dissertation thereon would occupy nearly as many pages as the whole of our treatise. We shall therefore content ourselves with pointing out the application of our preceding remarks, in conjunction with simple trigonometrical observations, and endeavour to elucidate the whole by some simple examples. In the case of a harbour or bay, not much enclosed with buildings, it is usual to survey the whole of it trigonometrically, *i. e.*, from a measured base, to take angles to various commanding points from either extremity, which points are subsequently used to take angles from to other points, and so on, until all the most prominent features are determined, both of the outline, and of such internal parts as are considered necessary, such as islands, rocks, shoals, &c. The position of all these points, we have observed, is obtained by taking two angles from *off* a base line; but where the third angle of each of these triangles is not observed, a third intersection should be obtained from a third point on the base line, and all the subsequent determined points should have at least three intersecting angles. When the most prominent points are thus determined, it is usual either to sketch in by hand the intermediate portions, or run lines by means of the azimuth compass—chaining or taking the distance in the usual manner. To lay down such a survey, it is necessary to calculate the sides of the principal triangles, which, as we have before observed, can be more correctly laid down by their sides than by their angles. The sides of the smaller triangles might be laid down by their angles, and, from there being three or more angles intersecting in the same point, a fair check will be obtained on its accuracy of position.

In the case of a town being on one side of a harbour, and open country on the other side, it will be proper to *base* the operations of the survey on the opposite side from the town, and, by carefully observed angles, determine certain positions on the town side, to which the instruments might then be removed, and the whole town included, if necessary, in the survey, either in the manner we have previously directed, or by a traverse.

A river may be surveyed with very little trouble, for, by running lines on both sides, and leaving marks at their extremities, or marks opposite to each other, angles might be taken across the river both ways, and be so effectually checked, that the distance across would be more correctly obtained than if it were possible to subject it to direct measurement.

In all these cases it is necessary to obtain the high and low spring line, and also the high and low neap line—which, without numerous sections are taken (in which case, from having the profile of the ground, and the height to which the tide rises, it could be correctly laid down on the plan), it will be necessary to measure it up by a chain or tape line, as in common surveying, or it might be determined by observing numerous angles from several well-determined points.

Supposing the boundary of the harbour or bay to be completed, it then becomes necessary to obtain the least depth of water, so as to be able correctly to lay down on the plan, the shoals, channels, &c., and the leading marks by which they may be determined; the respective depths being invariably reduced to ordinary *low* water spring tides. We have before remarked, that it is usual to determine the positions of small islands, rocks, &c., by the intersection of angles from ashore, and if a natural sharply defined point does not exist, to which observations can be taken, a mark must be erected. It is also usual with *land* surveyors, when employed in surveying harbours, &c., to explore and mark shoals, &c., with flag poles, and then determine their position by observing angles from the extremities of a base, or from other determined points on shore. This method, to a person not accustomed to the motion of a vessel, is perhaps preferable to any other, as the angles can be taken on shore to the required marks with more deliberation and certainty, than from a boat moored over the object to the points on shore; but in such cases the observer, in place of observing all his angles from one point, as when taken from the boat, has to observe them from at least three different points. It frequently happens that, from the indentation of cliffs, or from other causes, the boundary of a harbour or bay cannot be obtained by observations from a base line *on shore*, when, at the same time, if a base line could be laid down in the bay by floating marks, observations might be taken to points on it, from every part on shore. In such cases it is usual first to determine the length of a base on shore with great care, and, when such is done, to moor three buoys, somewhat in the shape of an equilateral triangle, and in a favourable position for being seen from every part of the bounding line. The distances of these buoys then, the one from the other, might be obtained by protracting the several angles observed from the base line on shore; but still, in such an operation, it would be only prudent to calculate each side of the triangle, and its position, from the observed angles and measured base. Afterwards, in filling in the boundary, it will also be necessary to calculate the distances of the most prominent points of the boundary, as a check on their protraction; but generally a careful protraction will be sufficiently accurate in determining all the minor points of detail. When sufficient time cannot be spared to measure a base line on shore, or when the nature of the country is such as to present insuperable obstacles to direct measurement, the length is sometimes computed by the velocity of sound; and so also is the distance of buoys or boats the one from the other, moored on a floating base line. Sound, according to Sir John Herschel, travels at the rate of 1090 feet per second at 32° Fahrenheit, and every additional degree of temperature adds 1·14 feet to the velocity; therefore at 62°—which is the standard temperature of the British islands—sound travels at the rate of 1124·20 feet per second. The most common method practised in obtaining the distance by velocity of sound is, to note the time by a seconds watch, between observing the flash and hearing the report of a gun; and when a great number of experiments are made, and the mean time taken, it is most probable a near approximation to the correct distance will be arrived at. The greatest care will, however, be necessary in conducting the experiments; and simultaneous or alternate experiments should be made from either station to counteract the effect of the wind, which in one direction would accelerate, and in the other retard the velocity; and it will be absolutely necessary, in arriving at anything like accuracy, that the minutest fraction of a second should be noted, as an error of only a single tenth will be equivalent to about 112 feet of distance; and it will of course be perceived, that the greater the distance of the stations, the more correct the results will be.

Soundings, in surveying, are usually taken with a lead-line differently marked to that used on common shipboard, but it may be marked in feet, yards, or fathoms, in any way considered best by the surveyor.* The lead-line employed by government surveyors is marked thus:—two and three feet, leather; five feet, white; seven feet, red; ten feet, two pieces of leather; thirteen feet, blue; fifteen feet, white; seventeen feet, red; twenty feet, two knots; twenty-two, thirty-two, forty-two, fifty-two, each by two pieces of leather; twenty-three, thirty-three, forty-three, fifty-three, each by three pieces of leather; twenty-five, thirty-five, forty-five, fifty-five, each by white marks; twenty-seven, thirty-seven, forty-seven, fifty-seven, each by red marks: thirty feet is denoted by three knots, forty feet by four knots, fifty feet by five knots, and sixty feet by six knots; and beyond ten fathoms, or sixty feet, soundings are taken in fathoms. The line employed is four-stranded, and the coloured marks are merely strips of buntin passed between the strands and knotted; the lead weighs from eight to twelve pounds, with an iron shoe at the bottom in the shape of a hollow cone, into which the *arming* of tallow is stuffed, for the purpose of bringing up a portion of the bottom when it strikes the ground, by which means its nature may be stated. By such information the ma-

* In shallow still water, or in slack streams, a pole, weighted at one end with a broad shoe to prevent its sinking into the bottom, and graduated in feet, may be very advantageously employed in taking soundings; and where the surveyor is not accustomed to the marks on a lead-line, perhaps the safest way would be to attach small tin plates, with the depth painted in white on a black ground, at the requisite distances.

riner is often enabled to tell his position on approaching the coast when all other means fail him; it is also of the greatest service in regard to anchorage and in grounding a vessel, to know the precise nature of the bottom.

The depth of water laid down in charts, we have before observed, is invariably that of ordinary low water spring tides, but as it will be necessary in every case to work at all times of tide, and at full springs as well as dead neaps, it will be apparent that a scale of reduction from the actual soundings taken must be prepared, to be used according to the time and state of the tides. The method most commonly practised in making these corrections is, to deduct three quarters of the *full* rise of the tide from all soundings taken during the first quarter ebb; one half the *full* rise of the tide from soundings taken during half ebb; and one quarter of the *full* rise of the tide from soundings taken during the last quarter ebb;* but it is generally considered dangerous to take soundings during the first quarter ebb, from the great variation frequently

* Mackenzie, in his work, gives the following rule, as sufficiently accurate for all ordinary operations:—

Spring tide at the			
	1st hour before and after high water, deduct ..	$\frac{11}{12}$ths.	of the full rise of ordinary spring tides.
	2nd	$\frac{3}{4}$ths.	
	3rd	$\frac{1}{2}$	
	4th	$\frac{1}{4}$th.	
	5th	$\frac{1}{12}$th.	
	6th	0	

Neap tide at the			
	1st hour before and after high water, deduct ..	$\frac{4}{5}$ths.	of the full rise of spring tides.
	2nd	$\frac{3}{4}$ths.	
	3rd	$\frac{1}{2}$	
	4th	$\frac{1}{4}$th.	
	5th	$\frac{1}{5}$th.	
	6th	$\frac{1}{5}$th.	

If the depth is taken at high water spring tide, deduct the full rise of ordinary spring tide from the depth found, and the remainder will be the depth to be inserted into the draught.

If the depth is taken at high water neap tide, from each depth deduct $\frac{4}{5}$ths of the full rise of spring tide, or the whole rise except $\frac{1}{5}$th.

experienced in its fall. It will be necessary, in order to make these reductions with accuracy, that every possible care and exertion should be made in obtaining the rise of the tides at different times, which should be reckoned, not from the low water of each particular tide, but from the low water of ordinary spring tides, which must first be carefully determined as the zero. It will, however, be a much better and more correct method to employ a man to watch a tide scale graduated from the previous determined zero, on each day when soundings are taken, by which means any irregularity produced by the position of the heavenly bodies, direction of the wind, floods, or other causes, would be neutralized; or, instead of employing a man for such purpose, a self-registering tide gauge might be used with great advantage. The government surveyors, and many civilians, when employed in taking soundings, are very exact in reducing the depth of cast to the level of ordinary low spring tides, by employing a man to watch a scale, who notes the exact time of every three inches rise or fall; the surveyor also notes the time of each sounding;—the depth of water at the scale, deducted from a sounding taken at corresponding times, will give the correct depth at low water.*

Having said so much relative to the taking of soundings, we will now proceed to explain the more difficult problem of fixing their positions. Casts of the lead are taken at certain intervals of time—regulated according to the required degree of accuracy—in a direct line from one point to another, and in other cases parallel to the line of

* It should always be carefully borne in mind that too little water be marked rather than too much—the former would be of little consequence in comparison with the latter, which, if not strictly attended to, might occasion the most serious consequences. Where there is the slightest shadow of a doubt in the surveyor's mind as to the precise depth of water, the least result should be recorded.

coast. It is scarcely ever necessary to lay down every cast of the lead—at least when they are taken within short intervals of time, such as the third or fourth stroke of the oars—and of course angles to fix the position of the boat are not requisite when the sounding is not intended to be plotted. The points where the depth of water changes are those only which require determining, and the intermediate spaces may then, if considered necessary, be divided by the number of casts, and laid down accordingly. In a case where buoys are not made use of in conducting the survey, it will be requisite, before commencing to sound, to determine several conspicuous points on shore, and if such do not exist naturally, flag poles are to be erected, whose position must be very carefully determined; and if extreme accuracy is aimed at, such marks should all be on the same level. When these are completed, it is usual to row from one mark direct to another, taking casts of the lead at certain intervals; and when a change in the depth of water takes place, one or two angles are taken to known objects, which fixes the position of the boat. One angle in such cases is sufficient to fix the position of the boat, but two angles renders the operation certain. When buoys are used in surveying the shore, great facility will be found in using them in their original positions in the after operation of sounding, and when such is not the case, it will be found a great saving of time, and the results be more accurate, to moor three buoys in a triangle, as directed for surveying the shore, and draw out lines in the direction of one of the sides, and sound along it as far as necessary—at every *particular* cast taking an angle to the other buoy, and a second angle, if considered necessary, to an object on shore. After sounding along the three produced sides of the triangle, lines should be taken *on* with one of the buoys and an object on shore whose position is known, but in a contrary direction to the latter, by which means *two* marks are constantly kept *on*, thereby insuring the straightness of the line; *particular* soundings are fixed by angles as before. Another method is, by mooring a single buoy in about the centre of a bay or harbour, and using several fixed marks on shore, *starring* off the soundings from the buoy in a contrary direction to the marks on shore, and fixing *particular* casts by angles, as previously described. This method may be easily understood by the aid of the following diagram:—

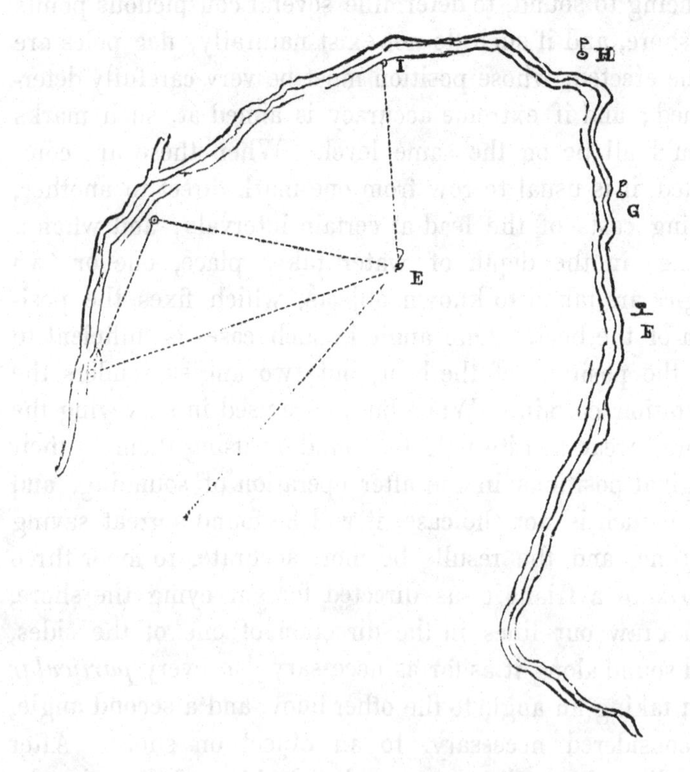

E is a buoy moored in about the centre of the bay; H G F are fixed marks on shore, by taking angles to which the position of the buoy E becomes known. Now, to commence a line of soundings—suppose from E to A—put the boat in a position so as to have the buoy E and the mark at G *on* with each other, and row in the direction of A; when arrived at the low water line, take angles to any fixed objects that may be visible, so that they do not lie very oblique to A G. The position of A being fixed, a line of soundings may be taken along shore by compass—without it can be got *on* with some fixed marks—until arriving at a point where some other fixed mark, as F, is *on* with the buoy, when row to it, sounding all the way; angles of course being taken at each change of water. When arrived at E, take a line on with E H, and fix its extremity in the same manner as A; but in taking a line such as E I, there will be great danger of swerving from the correct course, without another buoy is moored in the direction of I E produced, or two marks be got ashore which can be kept *on* throughout the line. In this manner soundings might be taken over every part of the bay with little trouble, and with certain results.

To plot the position of the boat when several angles have been observed from one point, it will be necessary to go into a trigonometrical calculation for determining the lengths of the sides of the triangles, and so be enabled to lay them down; or, otherwise, protract the observed angles on tracing paper from the central object, and draw out the several lines; lay the tracing paper on the map, and move it about until each line overlies the object to which the angles were taken, when the angular point will represent the position of the boat. A quicker method than this, however, presents itself in the station pointer—which is a kind of double segmental concentric protractor—having three projecting arms or rulers moving on the same centre, which can be set to form two angles, and the rulers being brought over the three points to which angles were taken, the centre on which they move will represent the position of the boat.* In concluding these remarks on the subject of marine surveying, we have only further to observe, that too much information cannot be embodied in such surveys; where leading sea marks are laid down, such as churches and other conspicuous buildings, it will be very proper, in addition to giving their correct ground plan, to sketch the elevation seawards, by which means no mistake can occur in regard to their recognition. The set and rapidity of the tides at various stages of flood and ebb should be carefully watched and recorded, and its direction shown by straight or inflected arrows. A sketch, showing the appearance of a harbour or bay from the offing, might also with great propriety be appended to such surveys.

In connection with marine surveying, we think a description of the buoys employed by the most eminent surveyors will be valuable to persons engaged in such operations. If a buoy is moored with a common grapnel and line, it is evident that sufficient line must be allowed for the greatest height to which the tide rises, otherwise it will be submerged at high water; but when such an allowance is made, at ebb tide the buoy will change its position, and continue to vary during every state of the tide, and only at extreme high water will it be in correct position. An ingenious contrivance has been adopted by some eminent government surveyors, as detailed in the annexed diagrams, which completely obviates the difficulty; but we will first describe an improved buoy for surveying purposes.

* A full description of this instrument will be found in Simm's Treatise on Mathematical Instruments, a most useful book, and which every engineer and surveyor should have by him.

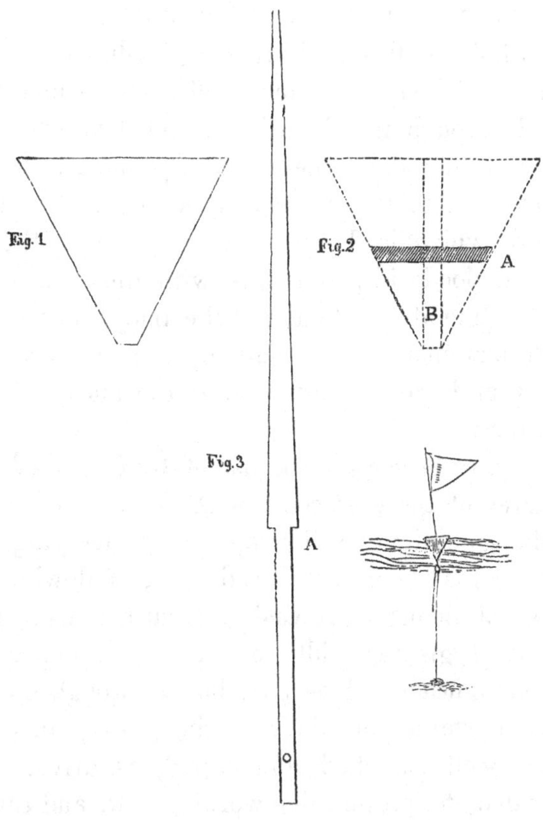

Fig. 1 is the buoy in the shape of a cone, and made of tin; it is about fourteen inches in diameter at the top, and twelve inches deep to the apex. Down the centre is fixed a cylinder of tin for the mast carrying the flag to pass through (see B, fig. 2); within the cone is fixed a thin piece of wood, with chamfered edges, to strengthen the sides of the buoy against blows, which of course is perforated in the centre to allow the tin tube to pass through it; the interior of the buoy is well varnished before closing, to prevent leakage. The mast (see fig. 3) is five feet in total length, and eighteen inches from shoulder to heel, projecting six inches below the apex of the cone, its shoulder, A, resting on the top of the buoy. Through a hole in this projection a play-block is lashed, as shown in the small sketch, and a three-quarter-inch line passed through it, which is connected at one end to a weight weighing about one hundred pounds, which rests on the ground; to the other end is attached a twenty pound lead, which is drawn upwards through the play-block by the flood tide, and falls with the ebb, keeping the mooring line always taut, and the buoy, consequently, erect and stationary. A small burgee, or triangular-shaped flag of buntin, is attached to the mast, to render it conspicuous.

The buoys or beacons made use of by Captain Belcher in the survey of the west coast of Africa are very simple, and, at the same time, well adapted for surveying operations, whether at home or abroad. The following is his description of them:—"A cask is furnished with double staves, and of greater width at bung and opposite, as well as double heads. Holes are bored through the bung and opposite staves, to admit of the passage of a spar. This spar, well parcelled and tarred, is driven firmly home, cleated, to prevent its working out, and caulked. Close to the cask at *b* a thimble is secured, through which the mooring cable passes, as well as one similar at *d*, where the cable is clove-hitched, parcelled, and its parts seized to the spar to prevent chafe. At *c*, sufficient ballast is attached to keep it steady. They were moored taut, and maintained an erect position in very strong currents and

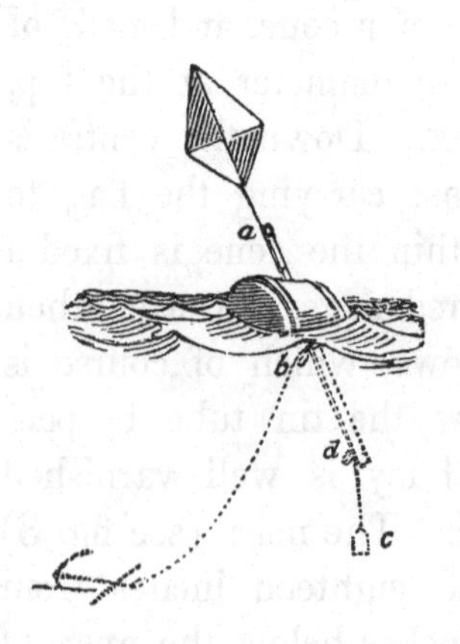

tides, and stood bad weather well." It will be perceived that the arrangement of the tackle for keeping the buoy steady over its mooring, without being influenced by the tide, is the same as that just previously described.

Inaccessible distances, such as usually occur in surveying operations, may generally be determined by practical geometry, with the chain on the ground, but such accurate results will not be obtained as when angular instruments are employed. In such simple cases of obstruction as trees, small buildings, &c., the method pointed out at p. 38 might with propriety be adopted, where also the lines e and f could be extended on both sides of a b; or, otherwise, the distance might be determined by merely setting out four right angles, as in the following example:—F *i* is a line passing through a building, and, on measuring up to it at *f*, set off a perpendicular, *f g*; from *g* set off another perpendicular, *g h*; from *h* another, *h i*; and from *i* another, *i e*; which will not only give the forward direction

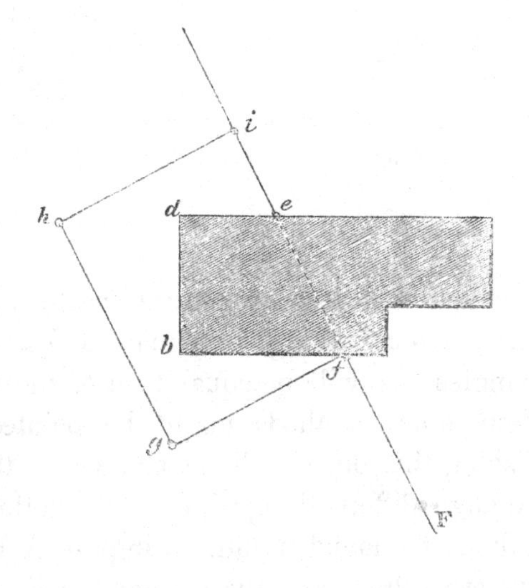

of the line, but, by deducting *i e* from *h g*, the diagonal distance through the building (*e f*) will be obtained. This method is, however, not to be recommended, for, as we have observed in our remarks on town surveying, an error in the first angle will be multiplied at the second, and so on for each; a saving of an angle would, therefore, be an important object.

Another method is by similar triangles: Thus—let *a d* be the direction of a line under measurement, the further progress of which is stopped by a pit, pond, house, or other obstruction:—From *a*, measure a line, *a c*, in any

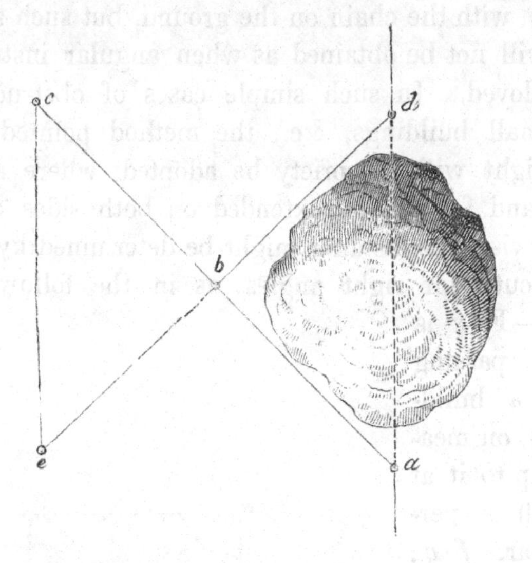

direction, and leave a central mark, *b*; from *d* measure a line, *d b e*, making *b e* equal to *b d*; then, by similar triangles, *c e* will be equal to *a b*, the required distance. Many other methods might be pointed out for accomplishing this object, all depending on the same problem, although differently applied. The following method may perhaps be found useful. Suppose A B to be a base or any other line, and the distance, *a* B, across the river to be required. In such a case, erect a mark at B, in the required direction, and another at the end of the measurement at *a*; measure a line, *a c*, at any angle with the base, and leave a *central* mark, *b*. Then measure a line, *c d*, parallel to A B, and the distance, *d*, where a line, B b, intersects it, will be equal to the required distance, a B. The great difficulty — it will be perceived by this method, lies in the setting out of the two lines B a and c d precisely parallel; we think the method given in the next page for determining an inaccessible station to be best for this purpose also.

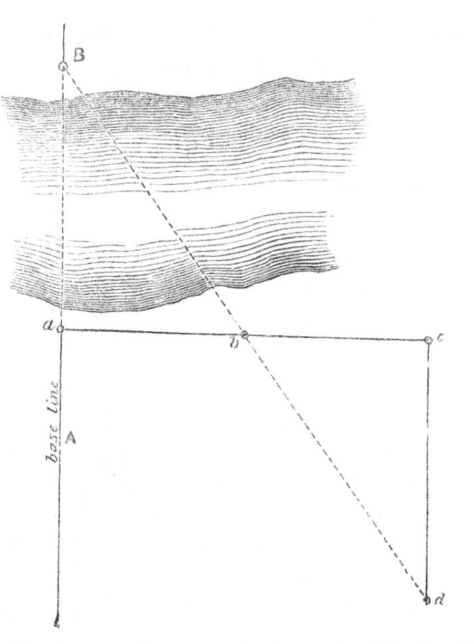

The accompanying diagram also presents a method of determining the distance across a river, but can scarcely be recommended in practice:—D A is a line to be continued across the river; the distance, A E, is consequently required. Leave marks, D and A, and also set up a mark at E, in its direction; measure any triangle, A B D, and also any other triangle, B C D; the vertex, C, of the latter being in the direction of the mark E, and of the vertex, B, of the former triangle; the distance, A E, may then be determined by scale, after plotting the lines on the plan; or, if A B and D C are set out parallel to each other, the

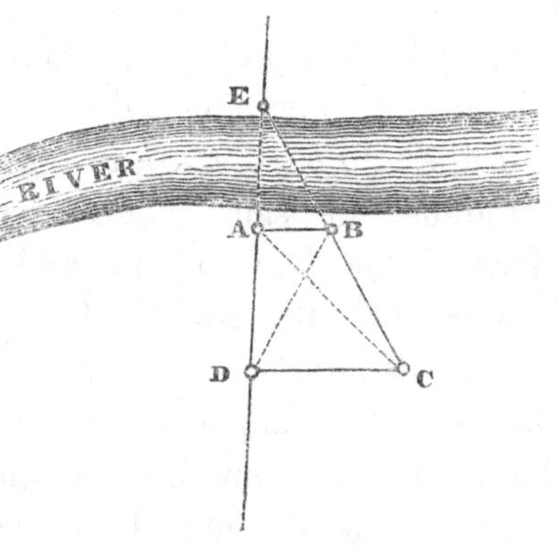

required distance might be calculated; for D C − A B is to D A as A B to A E.

To determine the position, on *both* sides of a building, of a line passing obliquely through it, such as *a f* in the accompanying diagram, continue the line up to *b*, where

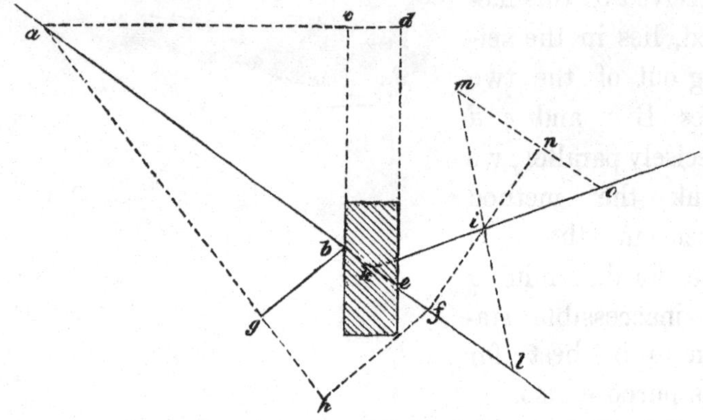

leave a mark, and take a new line right or left of it, as *a h*, and when arrived at about *g*, measure a perpendicular line to *b*, and set off another perpendicular line from *h*, so as to be clear of the building; the length of this last line, and consequently the point intersected on the other side, may then be obtained by a simple calculation:—thus, as *a g* is to *g b*, so is *a h* to *h f*. Another method of determining the same point may be arrived at thus:—Measure the distance, *a c* and *a d*, and also the perpendicular, *c b*; the length of a perpendicular, *d e*, would then be determined in the same manner as the preceding. Occasionally it happens in surveying, that, when a main line intersects a building, another line will close at some point on the previous line, occupied by the building, and which would be indeterminable by the ordinary method of direct measurement. In such cases, the distance on each line to the point of intersection may be ascertained thus (see last diagram):—Suppose the line *o k* to close on the line *a l*, at *k*—which point is evidently inaccessible—it will be necessary to determine the point *i* by direct measurement, through which measure a line, *f n*, making *i n* equal to *f i*; and also measure another line, *l m*, making *i m* equal to *l i*; produce a line, *m n*, to *o*, and *i o* will equal *i k*, and *n o* will equal *f k*.

The following method of mechanically determining the diagonal distance through a building is often found useful; but it will be perceived that the points on both sides of it first require to be determined. In the accompanying diagram, the line F E passes obliquely through the building, *a b c d*, and it is required to determine the exact distance, *f e*:—Take the *longest* side, *f b*, and lay it off on the other side, from *e* to *g*, observing that both sides must be laid off parallel to each other; then measure *b g*, which it is evident will be equal to *f e*. Sometimes a lean-to, or other projection, will be found at the end of the building, in which case take a distance, *f i*, sufficient to clear the obstruction, and lay it off from *e* to *k*, and measure *i k*, which will also equal *f e*.

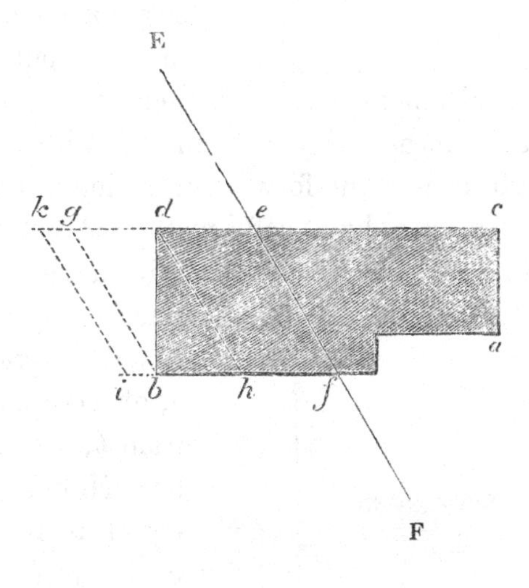

When angular instruments are used in conjunction with the chain for determining inaccessible distances, the operation will not only be much simplified, but more correct results be deduced, from the less circuitous process. For

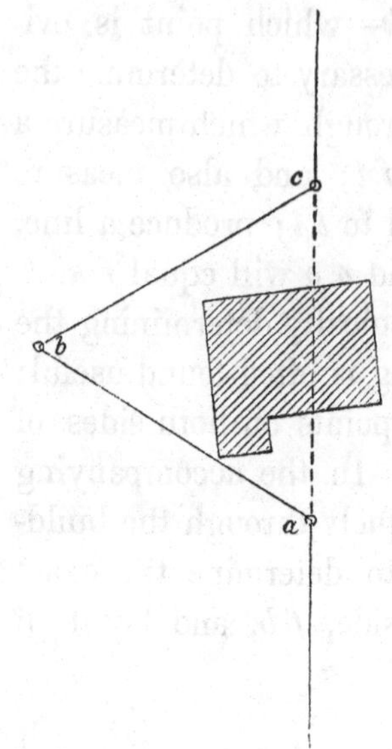

determining the diagonal distance through any obstruction, and the direction of the line forward—which latter particular is often of more service than the former—no method offers such facilities as the equilateral triangle:—Thus, in the following diagram, *a c* is the line obstructed in the direct measurement by the building; at *a* observe a *back* angle of 120°, and measure out any distance, *a b*; at *b* observe a *back* angle of 60°, and measure out a like distance, *b c*; the distance, *a c*, will then be equal to either of the other measured sides; and, by observing a *back* angle of 120° from *c*, the forward direction of the line will be given.

An exceeding useful method of determining the distance across a river presents itself in the isosceles triangle, as (incorrectly represented) in the annexed diagram:—From a point, *a*, observe any angle, *d a b*; and from *b*, or any convenient point on such line, take an angle, *a b c*, equal to half *b a d*; the point, *c*, intersected will then be the same distance from *a* as the point *b*. If *a b* is set off at right angles to *a c*, and the angular instrument planted at *c*, the distance across the river may be deduced, by observing any angle, and multiplying the distance in-

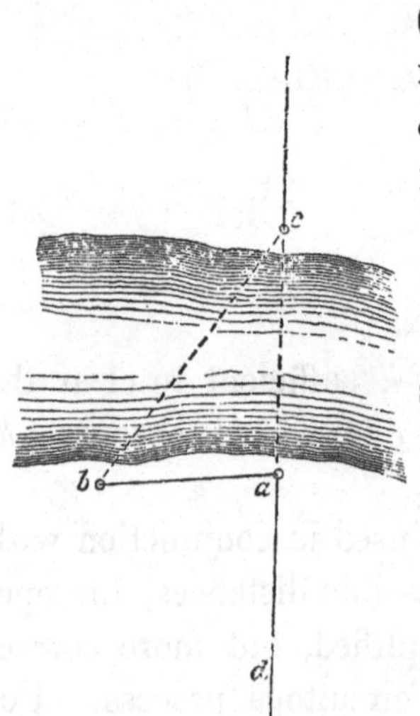

tersected, on the line *a b*, by the ratio of the base to the perpendicular; thus, if the angle *a c b* is observed to be 45°, the ratio will be one to one, and the perpendicular, consequently, be equal to the base; but if an angle of 26° 34′ be observed from *c*, the ratio of the base to the perpendicular will be as two to one, consequently the latter must be multiplied by two to give the required distance. The multipliers corresponding to various angles will be found in the Table of Slopes and Inclinations, at the end of the volume. The same result will be given if the complement of the angle is observed, and the perpendicular *divided* by the tabular number.*

We have proceeded thus far in our task of producing a manual of field operations, and, although we may possibly have omitted some few minor details, we trust that all which is really necessary and useful connected with surveying will be found embodied in the preceding pages.

The remaining part of the division on surveying will be devoted to the office, in plotting, copying, calculating contents, &c., in the description and adjustment of instruments, and in their uses in the field.

* Among the rude approximate methods of determining inaccessible distances—that which is practised by military engineers and also by some of the continental peasantry may be mentioned:—To determine the breadth of a river, they stand at its edge on one side and pull the rim of their hats over their eyes until the line of sight just cuts its edge and the other side; they then steadily turn half round, and observe where it cuts the ground, which they pace or measure for the inaccessible distance. Another method is to drive in a stake at the river's edge, and a second at a certain distance from it, and of such a height that, on looking over both, the line of sight cuts the opposite shore; the second stake is then pulled up and driven at a similar distance and height from the first, but in such a direction that the distance on the ground intersected by the line of sight can be measured. It will not fail to be perceived that in determining distances by this method the ground must be level, to arrive at even an approximate correct result.

CHAPTER VI.

ON PROTRACTING ANGLES. — PLOTTING, AND PLOTTING SCALES. — CALCULATION OF AREAS. — REDUCING AND COPYING OF PLANS. — OFFICE INSTRUMENTS AND FIELD INSTRUMENTS:—THEIR DESCRIPTION, USES, AND ADJUSTMENTS. — CONCLUDING REMARKS.

WE have previously remarked, that it is usual with practical surveyors to plot their work daily, as it advances, which, if possible, should be an invariable rule; otherwise, if delayed for a few days, and an error should have been committed in the first day's work, it will be very troublesome to correct, and incur a great loss of time; but, on the contrary, by plotting daily, any mistake that may have been committed will be easily rectified. A further advantage to be derived from pursuing this course will be, the scrupulous exactness with which every part of the work will be laid down, from every part of the survey being fresh in the operator's mind; and in all surveys there are particular points which, however elaborate the field-book may be, will still require the aid of memory before it can be correctly represented; for this reason it is always improper to allow a second party to plot a survey, however minute and straight-forward the system by which the field-book is kept. If it should be found inconvenient to *plot* daily, the lines of construction should certainly be laid down; the method of doing which, where a chain only is used, has been pointed out in a former part of our work; and where a magnetic angular instrument is used, directions will be found in the description and use of the circumferenter.

The method of laying down a survey made with a theodolite or sextant will, therefore, only be given here; to do which, with any degree of accuracy, a circular metallic protractor is indispensable. This instrument, for the general purposes of surveying, should be of about five or six inches diameter, divided on silver in a similar manner to a theodolite, with two projecting arms carrying verniers, and a third, by which the other two are moved round the circle, either with a rack and pinion, or clamp and tangent screws; but where great accuracy is required, the latter is preferable.* The projecting arms carrying the verniers have each a branch, with a fine pricker at its extremity. The inner part of the circle is chamfered off to an edge on opposite parts of the circle—or 180° apart—or sometimes at each quadrant, and the divisions brought down to it. A small circular space of metal in the centre of the instrument is removed, and a circular disc of glass inserted in its place, on which are drawn lines crossing each other at right angles, and dividing the small circle into four quadrants, the intersection of the lines denoting the centre of the protractor. When this instrument is used for laying down an angle, it must be so placed on the paper that its centre exactly coincides with, or covers, the angular point, which may easily be done, as the paper can be seen through the glass centre-piece. The divisions at 360° and 180°—which are brought down on the internal chamfered edge—must be on the line passing through the exact spot over which is placed the centre of the instrument. When the pro-

* We have had a protractor constructed with a rack and pinion, and also a clamp and tangent screw, which we find to possess decided advantages over those of the usual construction, particularly in the absence of all liability to shift its position when laying off several angles from one point, which is a common occurrence with those instruments having only the clamp and tangent screws.

tractor is thus placed, it is prevented from moving by four small studs, which take sufficient hold of the paper without damaging it; then, by means of the rack and pinion, or clamp and tangent screws, the vernier may be set to the required angle. A slight downward pressure on the extremities of the branches will make two small punctures in the paper, a line passing through one of them and the angular point or centre will be the required angle. The use of the second vernier is to check the result of the first, as often, in setting the instrument to the required angle, the protractor is stirred, and its centre no longer remains over the angular point. When such is the case, a line drawn from their punctures will not pass through the centre: the branches will also sometimes get deranged, and the same consequences ensue. To correct this, the branches must be altered by means of two small screws, on which they play, until a line will pass through the three points; this should always be attended to before the instrument is used in laying down the angles. When the angles on a survey are taken with a sextant, they are often laid down with a semicircular protractor without a vernier,* which may be also used with advantage when plotting a survey made with a theodolite, except for the principal angles, which should be laid down with the greatest possible accuracy. It should also be an invariable custom with a surveyor to protract all his lines before commencing to plot the fences, &c., as it often happens, on *closing* a day's work,

* In common semicircular protractors, great difficulty is experienced in laying off angles correctly from the oblique direction in which it is viewed—the light falling directly on it; but in a semicircular protractor which we have had constructed, the graduations on the chamfered edge are *within* the periphery, and consequently in shade, by which means the angle can be correctly laid off without difficulty. A further advantage of this plan is, the protection of the graduations from injury by accidental violence.

that the last line will not protract, which may arise either from some slight error in laying down the previous angles or distances, or in noting them in the field; if such should be the case, and the early part of the work plotted, it will be so much waste of time.

Plotting a survey does not comprehend the laying down of the lines of construction, but simply marking off the distances on those lines at which objects are intersected, or from which offsets to objects are taken, the lengths of those offsets, and the delineation of the various features to which such offsets have been measured. Offsets are usually taken at right angles to the line measured from, but sometimes obliquely in the direction of the fence. Where this is the case, of course the direction of the fence must be previously fixed, or otherwise the angle at which the offset is taken be measured. A common method of procedure in plotting is, first to mark off the distances on the line, and then, with the same scale, the several offsets determining the distance and position of the required object; of course applying the scale at right angles to the main line. Where there are many offsets, the distances should be written opposite a few of the marks, so that, in the event of a mistake, little progress will be made in the plotting before the mistake will be detected. Another method in common practice is, to lay the edge of the plotting scale on the line, with a weight at each end, or have it pinned down, and move a short scale—about three inches in length—of the same ratio as the other, at right angles along its edge, and when against any distance where an offset was taken, mark off the offset without previously marking the distance, as by the former method. There is also another contrivance for this purpose, first used, we believe, on the Ordnance survey of Ireland, in which the offset scale traverses a groove in that which is laid on the several lines of construction, and when arrived at any point where an offset was taken, it can be marked off in one operation. After the lines of construction are all laid down, and previous to the commencement of plotting, it will be but prudent to apply the scale to the several distances at which false stations are put down, as errors will often arise in their first laying off, which will be detected on a comparison.

On the subject of plotting scales, we think little information is required beyond what is universally known, although it is not quite clear to us that surveyors are generally aware that box scales are much more correct to work from than ivory, and card-board or paper scales than either;* but a ready means of comparing the expansion or contraction of a scale with the work in hand presents itself, in simply laying off the scale on a spare place on the plan, and occasionally applying the scale thereto. Occasionally there will be found considerable discrepancies in plotting scales, when compared with each other, although made of the same material, probably arising from their being graduated when at unequal temperatures, although doubtless often the result of carelessness in the makers. The Tithe Commissioners in a measure remedy this, by stamping such scales (box only) as coincide with standard metal scales which they have had prepared; consequently surveyors have no excuse for using bad or improper scales, for their assurance lies in using such only as are stamped.

* The engine-divided card-board scales of Messrs. Holtzapffel are decidedly the best scales for surveyors' purposes :—First, for their accuracy of divisions; secondly, their extra length and cheapness; and, thirdly, their equal expansion or contraction with the paper on which the plan is being plotted.

The great art in computing the areas of irregular enclosures consists in the reducing of the boundaries into right lines, and then again into geometrical figures. A correct method of reducing or equalizing an irregular side or boundary to a mean line is explained in the annexed diagram:—Suppose the side of a field to be of the irregular form below;—draw a line A B, and at A

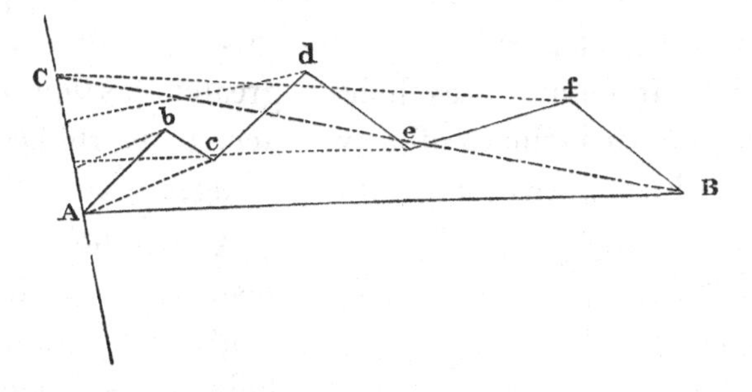

draw a transverse line, A C. Lay a parallel ruler from A to the *second* point or angular deflection at c; slide the ruler up to b, and draw the dotted line from thence to the transverse line, or, without drawing it, mark its intersection; from this last point lay the ruler to d, slide it down to c, and draw in the dotted line as before; from the point at which it cuts the transverse line, lay the ruler to e, and slide it up to d, and draw in the dotted line from d; lay the ruler from the point of bisection, to f, and slide it down to e, draw in this line, and, from the point of bisection, lay the ruler to B, slide it up to f, draw in the dotted line; and, from the point of bisection at C draw the line C B, which will equalize the irregular boundary, as much being cut off as taken in. This method is, however, rarely adopted in practice, too much time being taken up in the operation, and equally as accurate results being arrived at by a much shorter process, which is, to equalize those irregular boundaries by the eye, and by a little practice it may be done with the greatest exactness. A thin piece of transparent horn, or a strip of glass, is recommended for this purpose, by which means one may very exactly judge if as much new space is included as is excluded of the original; or a bow of whalebone, or any elastic substance, strung with horse-hair, will suit as well. But the method adopted by most surveyors is, to draw an equalizing line,—in pencil,—with a parallel ruler or straight-edge, on removal of which, if found to exclude a greater portion of the original than it includes of new space, it is rubbed out, and fresh lines drawn, until the eye judges it correct.

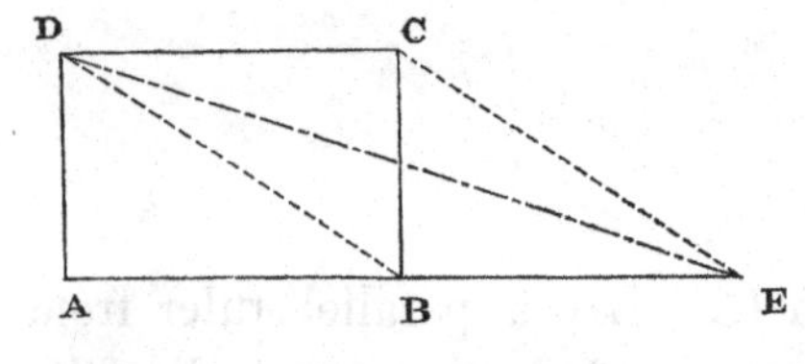

A parallelogram is easily reduced to a triangle of equal area, as follows:—Produce or extend one of the sides, A B, lay a parallel ruler on the diagonal, D B, and move it parallel until it cuts C; then draw in the line C E, or mark where it cuts the produced line at E; draw in D E, and it will be done: it will be the same thing if the distance A B is pricked off on the produced line, which will reach to E. The same method is pursued in reducing a figure of five sides to a triangle of equal area, as represented in the following diagram:—Let A B C D E be the given figure; extend the base A E on each side; draw lines C A, and C E, and also other lines, parallel thereto, to cut the base, as B G, and D F. The triangle G C F will then be of equal area with the given polygon.

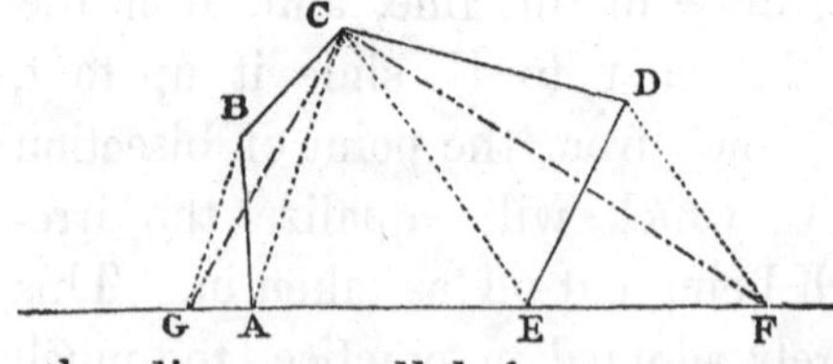

In computing the contents of any piece of land, whether it be one enclosure or a great number, it is usually done *quite independent of the several lines measured in the field*, except in some few cases where the base line and a few others, from their position on the plan, may be used with advantage; or, where extreme accuracy is necessary, the lines of construction and measured offsets are taken, and the area computed by considering each space between two offsets as a trapezoid, and the space between the lines of construction as trapeziums and triangles. In other cases, *new* lines are drawn on the plan, dividing each separate enclosure into trapeziums and triangles, the bases and perpendiculars of which are measured on the plan by means of the scale from which it was plotted, and then multiplied and added together for the total contents. After all the separate quantities are in such manner computed, and added together in one sum, the whole estate is calculated, independent of the fields, by dividing it into *large* triangles and trapeziums, and adding them together. If this last sum be equal to the former, or nearly so, the work may be considered right; but, if the sums have any considerable difference, there will be an error, and the whole must be examined and re-computed until the results nearly agree; or the contents may be found in a much more ready manner by dividing the whole estate into a great number of parallelograms, as hereafter explained, and afterwards proving it by large triangles and trapeziums, as above directed.

The area of any plane figure is the measure of the space contained within its extremes or bounds. This area, or the content of the plane figure, is estimated by the number of squares that may be contained in it;

the side of these squares being an inch, a foot, a yard, a chain, or any other fixed quantity; and hence the area or content is said to be so many square inches, feet, yards, chains, &c.

Land is estimated in acres, roods, and perches. An acre is equal to ten square chains, that is, ten chains in length, and one chain in breadth: also an acre is divided into four parts, called roods; and a rood into forty parts, called rods, perches, or poles. The chain generally used, called Gunter's chain,—from its inventor, the Rev. Edmund Gunter,—is four poles, or twenty-two yards, or sixty-six feet in length. It consists of one hundred equal links, and the length of each link is therefore $\frac{22}{100}$ of a yard, or $\frac{66}{100}$ of a foot, or 7·92 inches.

An acre of land then consists of

1,000 × 100 =	100,000	square	links.
660 × 66 =	43,560	,,	feet.
220 × 22 =	4,840	,,	yards.
40 × 4 =	160	,,	rods.

Lines measured with a chain are set down in links as integers, every chain in length being one hundred links; therefore, after the content is found, it will be in square links; then cut off five of the figures on the right hand for decimals, and the rest will be acres. These decimals are then multiplied by four for roods, and the decimals of these again by 40 for perches.*

* The great merit of Gunter's chain consists in his having so ingeniously availed himself of the accidental circumstance that the tenth part of an acre is equal to a square whose side is four perches or poles, or twenty-two yards. Hence he took twenty-two yards for the length of his chain, which he afterwards divided into one hundred parts, called links—a link being equal to 7·92 inches. Any decimal division of the chain would have answered his purpose, so far as the facility of calculation was concerned; but, had he divided it into ten parts only, there might be an error of 2·2 yards in every line. And had he divided it into one thousand parts, the numbers would be very large, and the inevitable errors of observation would far exceed the unit

The area of any parallelogram is found by multiplying the length by the perpendicular breadth or height, and the product will be the area required. The area of any triangle is found by multiplying the base by the perpendicular height, and half the product will be the area; or otherwise by multiplying one of those dimensions by half the other.

Example.—Required the area of a triangle, whose base is 1020 links, and perpendicular height 580 links.

```
   1020
    580
  -----
  81600
  5100
-------
2)591600
-------
2,95800
      4
-------
3,83200
     40
-------
33,28000
-------
```

```
   510
   580
  -----
  40800
 2550
 -----
2,95800
      4
-------
3,83200
     40
-------
33,28000
-------
```

Equal—2 acres, 3 roods, 33 rods.

To calculate the area of a triangle from the measurement of the sides, add all the three sides together and take half that sum; subtract each side severally from the

of the scale, which would be less than an inch. According to his division of the chain, lines may easily be measured nearly within half a foot, which is sufficiently accurate for all ordinary purposes of surveying. From what has been stated, it will be seen that an acre contains one hundred thousand square links; by setting off, therefore, five decimal places to the right from the results in square links—which is equivalent to dividing by one hundred thousand—we have the number of acres; whereas, if the same superficies had been obtained in square feet, we should, for the same purpose, have had to divide by forty-three thousand, five hundred and sixty—the number of square feet in an acre. We have heard many surveyors express their wonder why twenty-two yards was selected for the length of the chain!

said half sum, and multiply these three remainders and the half sum together, and extract the square root of the product for the area of the triangle. When the angle subtended by two sides is known, as well as the length of the sides, the area may be found thus:—multiply the two sides together, and take half their product; then as radius is to sine of given angle, so is the half product to the area: or multiply the half product by the natural sine of the angle, which will also give the area.*

The area of any trapezium is found by dividing the figure into two triangles by a diagonal, adding the perpendicular heights together, and multiplying by the diagonal; half the product will then be the area: or multiply one of these dimensions by half the other. In place of measuring the two perpendiculars separately, and adding them together, and then halving the sum, it is a preferable method to lay a parallel ruler upon the diagonal, as A B, and move it parallel to the angle, C or D, and draw the line C e or D f, then take off the distance with a pair of compasses from one of the angles to the nearest part of the dotted line, passing through the opposite angle, which will give the same distance as the sum of the two perpendiculars; apply this extent to a scale double that to which the figure is laid down by, and the trouble of adding and dividing will be saved. This method has also the advantage in point of correctness, for in measuring the perpendicular heights separately, there will

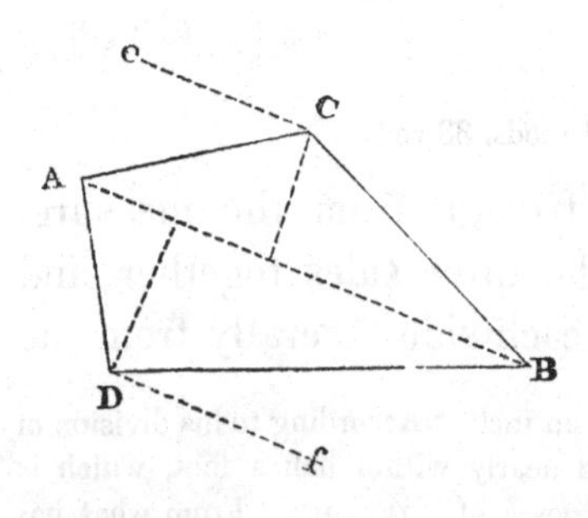

* When the content of an enclosure is found in square yards, it can be easily reduced to acres by multiplying by ·0002$\frac{1}{30}$.

be small fractional parts that cannot well be estimated, although their omission will make a sensible difference in the content.

The area of an irregular polygon* is found by drawing diagonals, dividing the proposed polygon into trapeziums and triangles. The areas of these figures are then separately found, and added together for the contents of the whole figure.

The area of a trapezoid is found by adding together the two parallel sides, multiplying their sum by the perpendicular breadth or distance between them, and half the product will be the area; or by multiplying one of those dimensions by half the other.

Example.—In a trapezoid, given—the parallel sides, 750 links and 1225 links, and the perpendicular distance between them, 1540 links; to find the area—

1225
750

1975 × 770 = 1,520,750 square links = 15 acres, 33 rods.

The area of any figure may also be found in an easy manner, by forming parallelograms on the plan. This is usually performed by drawing parallel lines, in some faint colour, all over the plan, forming squares of one or two chains each, which are afterwards numbered consecutively: the contents of all the squares are then easily ascertained, to which is added the contents of the broken squares, which are equalized, and calculated as triangles or parallelograms, and the area of each generally inserted on the plan. An example of this method of calculating areas is shown in plate 5. The squares are of one chain

* A polygon is ANY irregular figure having more than four sides, and, consequently, more than four right angles.

each, consequently the number (79) divided by 10 (the number of square chains in an acre), or, what is the same thing, cut off one figure on the right hand, which multiply for roods and perches—that on the left is acres, thus:—

Acres	7,9 the number of squares.
	4
Roods	3,6
	40
Perches	24,0

		A.	R.	P.
Contents of squares	=	7	3	24
Ditto broken squares	=	1	1	31
		9	1	15

It will be perceived on reference to the plan, that in some of the broken squares the contents of an entire square (16 perches) is inserted; but it will be noticed that immediately adjoining these broken squares, in which it is so inserted, that a piece equal to the deficiency is omitted in the calculation. Also, where a piece in a broken square is very small, it is carried into the contents of the adjoining one.

The great difficulty in computing areas correctly, lies in the probability, nay, almost certainty, of errors being committed in the numerous figures required to work out the contents of a large plan; and, to remedy this evil, numerous ingenious methods have been resorted to, either entirely to abolish computation by measurement, or to reduce it almost to a mechanical operation. Under the former head was a proposition—made, we believe, by the late Dr. Hutton—to cut to pieces the original map, or a carefully prepared copy, on the lines of the fences, and accurately weigh the pieces by a delicate balance;—and, of course, by first preparing tabulated results of such balance, or, in other words, the weight of any number of perches, roods, and acres of surface of the same paper as the plan—the measurement would be easily deduced, although we conceive not with very great accuracy. A much more feasible method was subsequently practised, by which the area of large maps have been computed with extreme accuracy. This plan consisted in affixing, by thin mucilage, a carefully prepared tracing of the map to be computed, on thin rolled lead, cutting the tracing, and consequently the attached lead, to pieces, weighing them as before, and comparing the weights with tabulated results. Or, otherwise, by preparing a quantity of pieces of lead, each representing an acre of surface; also other pieces, representing roods and perches; and a sufficiency of these being put into the opposite scale, so as to counterbalance that whose area was required, the content would be represented by such quantity. We believe that this plan has been tried at the Tithe Office, and that in very many instances the results so obtained were exceedingly accurate as compared with the ordinary, but tedious process of scaling; but incorrect results were often obtained—probably from the difficulty of obtaining lead of equal thickness and density—which have induced the authorities altogether to abandon the process. Another mechanical contrivance for computing the contents of irregular plane surfaces, is detailed in the Franklin Journal; it is the invention of a Dr. Wood, and is thus described:—"It consists of two plates of plain ground glass, with their inner surfaces fixed in a frame, so as to be parallel to each other, and only so far distant as to permit a piece of drawing paper to slide easily between them. They are of a rectangular form, fastened on three sides in any manner which shall leave the surfaces parallel, the fourth side being open; the space within is then partly filled with pure quicksilver, and, by means of a slip of drawing paper, the outer edge of the quicksilver is made straight and rectangular with the sides: its position is then marked. This may be done by noting on the paper used its distance from the outer and open edge of the glasses. The plot of any irregular surface is then moved in till the quicksilver extends to that point of the plot which is nearest the outer and open edge. The outer edge being now parallel to the former edge, by the manner in which the paper containing the plot is cut, its distance from its former edge is measured or marked on the same paper, and the area of the irregular field is thus found to be the difference of the areas of two given rectangles."

To abbreviate the labour and attendant inaccuracy of computing areas by the usual method of triangles, many instruments have been invented, the best of which we shall proceed to describe. The first which we remember to have seen applied to practice was the Land Surveyor's Calculator, by Mr. George Heald, which is constructed as follows:—Four concentric circles of cardboard, or other material, move about a centre, and a fifth, or outer circle, is fixed. The circumference of the latter is divided into 1000 logarithmic portions, representing links, and numbered at every ten divisions—10, 11, 12, &c.—which numbers represent respectively, 1 chain, 1 chain 10 links, 1 chain 20 links, &c.; and these divisions are again subdivided into links. The next circle is moveable, and is graduated on its edge in every respect similar to the preceding, but in a reverse direction. The second moveable circle represents acres and decimals, commencing with half an acre, and proceeding up to five acres. The next circle is divided into perches, to correspond with the preceding decimal division, which is numbered at every ten perches; and the acre is expressed at the termination of every rood. The last, or fifth circle, shows the area in acres, roods, and perches; when it is under two roods, it is numbered accordingly. The method of using this instrument is as follows:—Having the diagonal, and the two perpendiculars of a quadrilateral, or the base and perpendicular of a triangle, the areas of the respective figures are found at once, on inspection of the fourth or fifth circle, according as the area is greater or less than half an acre. There have been several instruments recently brought forward of a similar construction and purport, but none improving on the original.

The next instrument which we heard of was the Pediometer invented by Mr. F. Scheereck, for which the Lords of the Treasury granted him £100. The annexed diagram and description will fully explain this instrument. It consists of a square, *w a b*, and a graduated scale, constructed for three chains to an inch; *g* is a milled head, by turning which, motion is given to the brass slider, *b*, and the two pointers, *r* and *w*; *i* is the index to be placed in coincidence with the zero division upon the scale. When

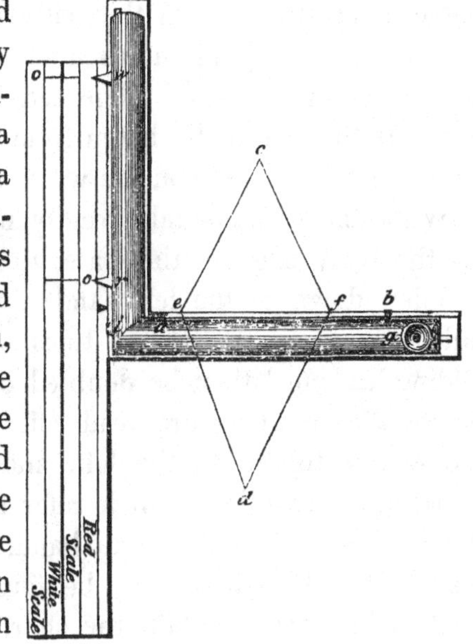

the brass slider, *b*, is in contact with *a*, and *i* coincides with the zero division, *r* and *w* at the same time pointing to *o* upon their respective divisions on the scale, the instrument will be in adjustment. When it is deranged, restore it by opening *r* and *w* to the proper distance, moving *a* into contact with *b*, and *i* into coincidence with the zero division. In the diagram the instrument is represented in use, the area of the trapezium, *e c f d*, being required. The edge *a* is placed upon the point *e*, and *b* is opened to the point *f*; the square is then held down firmly with the right hand, and the scale placed against the edge of it with the left. The scale is then held down firmly, and the square slid up until the edge *a b* is upon the point *c*; the square is then held down firmly, and the scale slid against its edge until zero coincides with *i*. Finally, the scale is pressed and the square slid down until the edge *a b* is upon the point *d*, and the numbers to which *w* and *r* point, taken out; the latter is then subtracted from the former, and the contents in acres and decimal parts of an acre is at once given. The red pointer directs to the numbers that are to be taken from the red scale, and the white ones to those upon the white scale.

When the pointers fall exactly upon the line engraved on the ivory edge of the scale, the folding leaf is to be doubled down to the left hand; but when the pointers fall between any two of the lines on the ivory edge, the folding leaf must then be doubled over to the right hand before the numbers are read off. For instance, when the leaf is turned to the left, and the red pointer falls between the two lines which refer to ·008 and ·013, turn the folding leaf to the right hand, and the pointer will read ·010. It will always be found the most accurate method in practice to take the shortest diagonal of a trapezium, contrary to the usual practice in common scaling operations.

The following simple process has been successfully employed by many surveyors in computing maps, and which depends on this property of a right-angled triangle, that if a perpendicular from the right angle be let fall on to the base, it will be a mean proportional between the segments of the base, and therefore equal to the square root of their product;*—thus, in the accompanying diagram:—

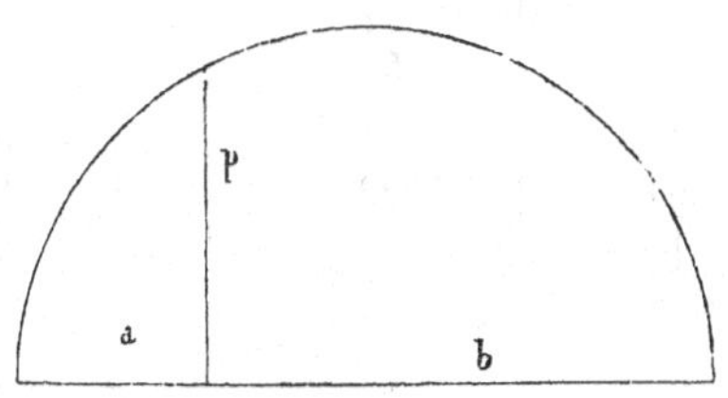

$$a : p :: p : b; \text{ therefore } p = \sqrt{ab}.$$

The practical application of this principle is as follows:—Extend the diagonal of a trapezium, or base of a triangle, and lay off their perpendiculars thereon; from the centre of this extended line describe a semicircle, and raise a perpendicular from the extremity of the diagonal, which perpendicular will be the line p, equal to the square root of the product of ab. To ascertain the acreage of any figure, the square root of whose area is thus found, take the squares of all numbers as high as necessary, and arrange them in a table as follows:—Column 1 contains the square roots; column 2, the half square, which is square links; and column 3, the contents, in acres,

* An angle within a semicircle is a right angle; and, supposing a triangle to be formed on the extremes of the base, a b, the line p, being drawn from the right angle, perpendicular to the base, the triangles on each side of the perpendicular would be equiangular, and similar to the whole triangle, and to each other, and, consequently the sides proportional.

roods, and perches, answering to the square links in column 2.

Square Roots.	Half Square (Square Links).	Contents.
1	·50	
2	2.	
3	4·50	
4	8.	
5	12·50	
|	|	
10	50.	
|	|	A. R. P.
100	5000.	0 0 8
|	|	
1000	500000.	5 0 0

The accompanying diagram, a b c d, represents an enclosure to be computed by this method. Produce the diagonal, a b, towards g; take off the perpendiculars, e d and f c, and lay them off from b towards g, to which

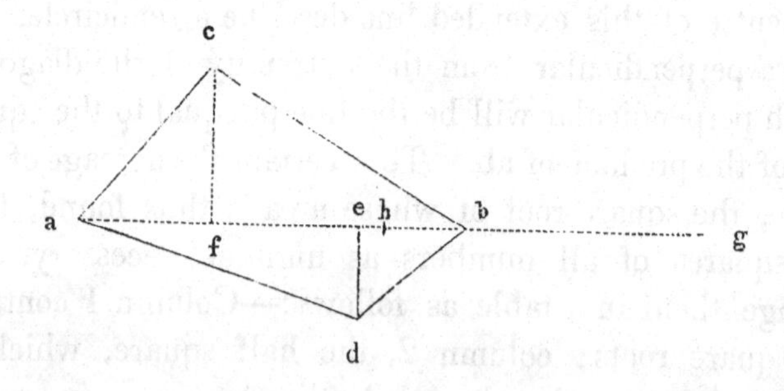

point they extend; and, with the instrument just described, find the centre of the line, a g, which will be at h; then, with the radius, h g or h a, put one foot of a pair of compasses in h, and with the other bisect a perpendicular raised from b, which will give the square root of the products of the two quantities—on whatever scale the diagram may be constructed—*viz.*, the diagonal, a b, and the sum of the heights, f c and e d, but which product requires to be halved to give the area; but with the assistance of the above table, the contents will be immediately given for the root of any product; or the product will be given without the trouble of halving, by the blade being double the scale of the plan.

Great facility will be derived, in the computing of maps by this plan, to have a simple instrument constructed, as the following:—*a b c* is a small ⊤ square, with the blade divided to the same scale as the plan is plotted to, and made to traverse the stock by a groove or other means, and capable of being fixed by means of a clamp screw at any part of the stock, the blade remaining perpendicular the while. The stock is simply divided into a scale of equal parts, commencing at the centre. The elongated diagonal is then easily halved.

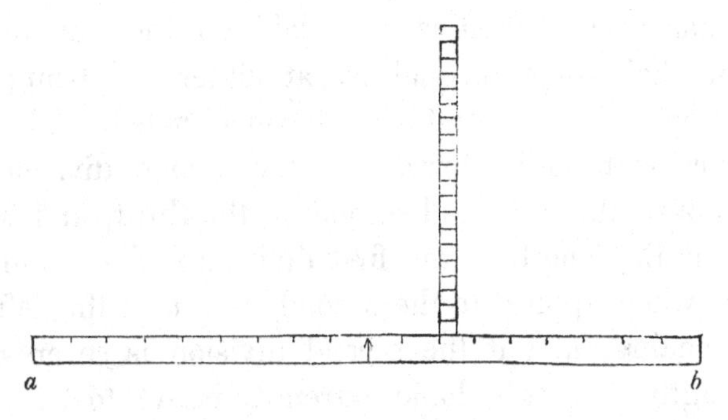

To use this instrument, the centre, h, is found by the scale on the stock, and the blade is moved along the stock until the fiducial edge is on the point b; the compasses being then placed with one foot in h—or the centre of the stock—and with the radius h a or h g, the other will intersect the scale on the blade at the required division, and which will be the square root, in links, of double the content.

The last, and most simple method, which we shall describe, and which is now in the course of adoption by all surveyors, and at present exclusively employed at the Tithe Commission Office, presents the greatest facility in performing computations, without in the least damaging the plans, by equalizing boundaries, &c., as by all the previous contrivances. The principle of the plan has long been in use by some few surveyors, but they prudently kept it to themselves, in order that the price of such work might not be reduced; but at last the method has become publicly known, and a vast reduction has taken place in the remuneration of such operations. In the first place, tracing paper of a superior quality is procured, and parallel lines, at exactly one chain apart, drawn in one direction only along the whole width of the paper. This paper is then carefully laid over the enclosure which is to be computed; the scale to which the map has been plotted is then laid on the first division of one chain—the inequalities at either end being equalized by the eye—and the distance noted. This first distance is brought forward at the second division, and the sum of the first and second at the third, and so on; thus, if the length of the first division is five chains, the scale, when applied to the second, is set on the left hand at 5 chains; and if the second division is seven chains in length, the right hand extremity is set to 12 chains, which quantity is again brought forward at the third division, and so on until the whole distance of a field, in strips of one chain, is ascertained, when the acreage is at once deduced, by cutting off three figures from the right hand—those on the left are acres—which are multiplied for roods and perches. An ingenious application of the above system is now in operation at the Tithe Office, by which means all calculation is avoided, and the area has merely to be read off on a scale. The following diagram and explanation will enable any surveyor instantly to practise it.

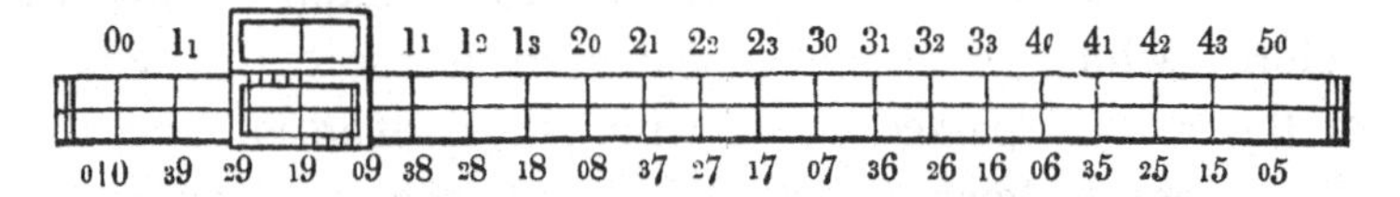

The instrument consists of a box rule, with divisions at $2\frac{1}{2}$ chains apart, and numbered 0_0, 0_1, &c.; at four of these divisions, or ten chains, it is numbered 1_0, or an acre—the reader bearing in mind that the divisions, on the tracing paper laid over the field to be computed, are one chain apart—therefore each single division, as 0_1, is a rood. There is a brass slider attached to the rule with a horsehair strained perpendicular to its length, for the purpose of equalizing the fences at the end of each strip. On this slider—which embraces rather more than two roods in its aperture—are laid off 40 divisions, on alternate sides, each way from the centre, and which are exactly the length of one rood, consequently each division is a perch. The figures on the upper side denote the acres and roods, as far as the rule extends, and are continued backwards on the lower part; the large figures are acres, and the small figures roods. Now, to apply this instrument to practice, lay the ruled tracing paper over the enclosure, and move the slider until its centre is on 0_0,; place the scale in such a position that the horsehair forms a mean line of such part of the left hand bounding fence as is included in the first strip of one chain wide, and press it gently on the paper; with the right hand move the slider along the rule, until the horsehair forms a mean line on the required part of the right hand bounding fence. Then move the instrument *altogether* on to the next division—the slider still remaining as last set—the horsehair forming a mean line, as before, with the left hand hedge; press the rule gently, and move the slider on the scale, until the horsehair forms a mean line with the right hand hedge, as before; which process is repeated until the entire length of the rule is passed over, when it is reversed, and the slider moved towards the left hand, the equalization in this case commencing first on the right hand side. When the slider is brought back to its original starting point, if there remains any further quantity, it is again moved forward to the right, as at first, the continuous measurement being easily kept up by the decimal arrangement of the contents. For example, in the position the slider occupies in the diagram—supposing it had been moved over the scale and back—the contents would be 10 acres and 3 roods; and if, instead of the centre exactly coinciding with the division representing 3 roods, it was 20 of the small divisions on the slider beyond it, the contents would be 10 acres, 3 roods, and 20 perches. As a proof of the great saving effected by this instrument, we need only observe, that the price of scaling has been reduced from fifty to seventy-five per cent. since its introduction.

On the subject of reducing and copying plans we cannot be expected to say much. For ordinary purposes the pentagraph presents the readiest method, both for copying on the original scale, and also for reducing or enlarging the copy to any proportional size of the original. There are, however, several improved instruments for copying with greater accuracy than the common pentagraph admits of. The reducing of a plan by hand, is commonly performed by drawing squares of a size commensurate with its minutiæ all over its extent. Similar squares of any required proportion to the first are then drawn on the paper on which the plan is to be copied, and in every square of the copy is constructed that which is contained in the corresponding square of the original; to enlarge a plan the operation is reversed.

A much more accurate method than the above for reducing or enlarging plans for railways or other similar purposes, is, to lay down lines of construction thereon, in precisely the same manner as would be done in surveying it; then take off the lengths, offsets, &c., with the proper scale, and replot the survey to that scale on which it is required. The usual method of copying plans by hand is to prick all the angular points and principal features through the original on to a plain sheet of paper fixed beneath it, on which the copy is to be drawn; these points being then connected—first with pencil lines—are inked in, and a tolerable accurate copy obtained; but the method is not to be recommended, from the injury it does to the original, and the incidental errors from oblique punctures of the pricker, &c. The best method of copying plans, which we are aware of, is either by a copying glass, or by tracing and transferring. That by the copying glass is performed thus:—in a frame, which can be fixed at any inclination, is placed a sheet of plate glass; to the frame is fixed the original plan, and above it the paper on to which it is to be copied; the frame is then placed behind a strong light—or lighted candles placed below it—which enables the draughtsman to see all the lines of the original, and to trace them in ink on the plain paper without difficulty.* The second method

* We have frequently practised this method in copying railway plans and sections in the country—using common window glass for the purpose, and found a great saving of time in comparison with the common method of pricking through.

is to make a tracing of the original on proper tracing paper; rub the back of it with powdered black lead, and fix it down carefully on to the paper on which the copy is to be made; then lightly trace all the lines with the end of a porcupine's quill, or other pointer which will trace fine lines, and a perfect copy similar to pencil will be obtained, which has then to be inked in.

The office instruments employed by the surveyor are few, and such as would only require a very brief description; but with one or two exceptions we shall content ourselves by merely referring the reader to a work published on the subject.* Of plotting scales we have already spoken—but have only incidentally referred to straight-edges, which is one of the most important instruments employed by the surveyor or engineer. All straight-edges we should recommend to be of steel, in preference to wood, brass, electrum, or any soft metal. In drawing a line exceeding the length of the straight-edge, at least half its length should be applied to that portion of the line already drawn, as a bearing by which the renewed or additional length might be correctly drawn in continuity of the same right line, and the same edge should be used throughout its whole length. A line would be drawn more accurately straight, if, at each repetition of the straight-edge, it was laid off alternately above and below the line, the same edge being used throughout. To test straight-edges—which in every case is indispensable—it is only necessary to draw a line by one of its edges, and turn it over, so that the same edge is presented to the line from above; by this means the slightest defect will be observed. We do not conceive that any fitter place presents itself than the present to

* A Treatise on the Principal Mathematical Drawing Instruments employed by the Engineer, Architect, and Surveyor: published by Mr. Weale.

offer a few remarks on the paper commonly used to plot surveys on. Mounted paper is always more affected by changes of temperature than that which has not been damped, and for very accurate purposes we should prefer the latter on that account. But for the common purposes of surveying such extreme accuracy is not requisite; and for the preservation of those documents it becomes necessary to have them mounted on strong linen; great care is, required in selecting a hard paper, and it should be kept for many months after mounting, to dry and season before use. These suggestions are, however, rarely attended to, and nothing is more common than for a surveyor to have paper mounted after determining the extent of a parish, and on which he will lay down his lines of construction within a few days of its preparation; and in many cases we have known surveyors have fresh mounted paper attached to seasoned paper where the original size has been too small, while the map was in process of construction, whereby we are satisfied great errors must have arisen. We have a map now before us, plotted while the mounted paper was green, which has shrunk, in a length of ten feet, more than one-third of an inch, equal to nearly seventy feet of the scale on which it was plotted. The comparison in this instance has been made with the same scale with which it was plotted, and in consequence of each repetition being pricked off throughout the distance, the contraction was at once perceptible. In making some additions to this map we have been compelled to adopt a scale derived from the map itself, otherwise the two portions would not correspond.

The Chain and Offset-staff are two of the principal instruments employed in surveying. The Chain, as before mentioned, is 66 feet, or 100 links, in length,* comprising ten divisions of ten links each; distinguishable by brass marks cut into fingers from one to four; one finger denoting 10 links; two, 20 links; three, 30 links; and four, 40 links. The chain is marked in this manner from both ends, the centre, or 50 links, being denoted by a round brass mark and swivel; it is therefore immaterial which way a line is measured with the chain. Four fingers beyond the round brass mark or centre denote 60, three fingers, 70, two fingers, 80, and one finger, 90 links: the end of the chain is 100. It will be found rather perplexing at first reckoning beyond 50, but a little practice will soon make it familiar; the student should, however, be very particular when reading the chain at about 60, or 70, as nothing is more common than the mistaking of these marks for 40, or 30.

The Offset-staff is a light rod, generally of 10 links in length; the links being marked by notches or brass nails, though, to a person at all accustomed to surveying, such divisions are superfluous; but a notch should be cut at one end, for the purpose of retaining the handle of the chain while it is being thrust through a hedge. The staff is carried by the person following the chain, which should be the surveyor himself, never trusting to a common assistant in this, the most particular part of his work. A tape is generally used for measuring buildings, and occasionally for taking offsets, although it is not so handy as the offset-staff, without an extra assistant is employed: a tape of 100 feet will be found the most advantageous length,†

* The surveyor's assistant should daily examine the chain, and, in addition to testing its length, straighten all the links, and close the rings where open. A pair of nippers, with a spare link and a few rings, should always be carried in the field, for the purpose of repairing the chain in case of accident.

† A self-acting tape in a metal case, (for which a patent has been obtained) is sometimes used by surveyors, which winds itself up by means of a spring,

and, in practice, care should be taken to hold the case vertical, to prevent the tape from chafe and quitting the reel.

The advantages of using a much longer chain than that of Gunter's are many—and which we shall proceed to point out, as well as the method of setting down the chainage and plotting the work. The reason for using a chain of 22 yards in length, in measuring lines, is not that it is the best length for securing accuracy, but from the facility with which the succeeding arithmetical computation of land is performed. The best length for a chain, for engineering purposes, is a desideratum yet undetermined,* although there can be no doubt Gunter's chain is not the most fit. We recommend, and, wherever it is possible, we use a chain of 100 feet in length, divided into feet; by which means the reduction to yards, in the computation of earth-work, is much more easily performed than when dimensions are taken on Gunter's chain.† When a chain of 100 feet in length is employed, the chainage is entered in the field-book, in the same manner as when the common link-chain is used; the only difference being in the plotting of the work, or rather in the graduating of the plotting scale. If a survey is made with a 100 feet chain, and plotted to a scale of 5 chains to an inch, the scale is divided into 330 to the inch, there being that number of feet in 5 chains; that is, there are 33 divisions to the inch, each division representing 10 feet. To plot to a scale of

but a much better instrument would be obtained by attaching a multiplying wheel to the axis or reel of the common tape. We have generally found that tapes permanently elongate about 3 inches in 50 feet, after a few days use.

* In an advanced part of our work will be found a description of a chain and levelling staff, for engineering purposes, divided, decimally, into yards, which presents great facility in the computation of earth-work.

† To turn links into feet, multiply by ·66, and to turn feet into links divide by ·66; but as the measuring tape is usually divided on one side into links, and on the reverse side into feet, it will answer the purposes of a table for this conversion, without any calculation.

$2\frac{1}{2}$ chains to an inch, there are the same number of divisions, each division representing 5 feet; but there are some scales into which the 100-feet chain cannot be so well subdivided, which will prevent its general adoption. Probably the most correct and *convenient* chain for general use is one of double the length of Gunter's, divided into 100 links, each link of which is double the usual length. This length may be used in the same manner as Gunter's; that is, the links set down as integers, each link counting but one, although it is actually two. But in plotting, by using a scale double the size of that to which the work is to be laid down, the results will be the same as if a chain and scale of the usual kind was used: thus,—if the survey is to be laid down to a scale of 4 chains to an inch, by plotting with a two-chain scale the results will be the same as if the common chain was used, and the work plotted to the proper scale. Also, by having stamped thereon a second division of figures, denoting the proper scale of 4 chains, the work may be plotted, and the contents calculated, with the same scale.* But our chief reason for recommending a chain different from Gunter's is, that a line can be measured with much greater accuracy with a long than a short chain, for the same reason that a line on paper can be laid off more correctly with a long scale (within limits) than with a short one, *viz.*, that so many lengths are not required to be laid off, and each repetition producing an error which, although small in itself, when multiplied by a large number, produces a very sensible quantity. Of this any one may easily satisfy himself by simply measuring out several lengths

* Scales of a greater length than a foot or eighteen inches, when of wood or ivory, are not to be recommended, from their sensible alteration in length, by change of temperature; but *total* distances may advantageously be laid off from brass or steel straight edges, which may easily be graduated (into inches only) for such purpose.

of a short scale, and afterwards testing the distance with a much longer scale, although many people may say a mean of errors will produce correct results. Also, in measuring over a wall, a bank, or a thick hedge, through which the chain cannot well be passed, the angle formed with the horizon by the long chain will be more obtuse, and approach nearer to a level line, than where a short chain is used. In the latter case a link, or several links, may be imperceptibly lost, and, if attempted to be rectified by drawing the chain a little forward, it will be at random, and consequently very uncertain. Rapidity in measuring, in reading the chain, and in plotting, are further advantages attached to the use of the foot, or double link chain; although for computing contents it will be found most convenient to use a scale of common chains. For all these reasons we recommend the use of a longer chain than that of Gunter's, especially to engineers.

The Vernier* is a contrivance for measuring the fractional parts of a scale, contained between two of its least divisions. The principle, and application of this fractional scale to angular instruments, may be easily understood:— suppose the lower plate of a theodolite to be divided into half degrees or thirty minutes, and that the space occupied by thirty-one or twenty-nine of such divisions is taken to form the scale on the upper plate called the vernier, and is divided into thirty equal parts, the difference of one division on the vernier and lower plate will be by excess or defect $\frac{1}{30}$th of the primary scale, or a single minute; and the difference of 2, 3, &c., divisions will be 2, 3, &c., minutes, and at the thirtieth division on the vernier, or

* So called from its inventor, Peter Vernier, a Frenchman, of the seventeenth century. The vernier is often termed "a nonius," from Nonius, a Portuguese mathematician, who published an account of an inferior instrument of this kind in the sixteenth century.

twenty-ninth on the plate, the difference will of course be 30 minutes, or half a degree, by excess or defect. By this means, any fractional part of a degree, in minutes, is at once seen, for when the zero of the vernier is between two of the divisions on the plate, look along the vernier until one of the divisions coincide with some division on the plate—the number of divisions from the zero at which such coincidence occurs will be so many minutes to be added to the half degree into which we suppose the plate to be divided. In like manner, if the plate was divided into twenty minutes, to read single minutes by the vernier, it will only be necessary to take twenty-one or nineteen of the divisions of the plate, and divide the whole length into twenty parts for the vernier. When reading off the fractional parts of an angle, it is a good method to judge of its accuracy by the eye. Thus, if the plate is divided into half degrees, and 10′ is read off, estimate the distance past the last division, which would appear about one-third, if correct; in this manner, we have sometimes amended ourselves when entering a wrong angle.

We now proceed to the subject of angular instruments, commencing with the Circumferenter. This instrument is principally used in mines and coal pits, and in surveying a thickly wooded country. Each angle is taken with the needle from the meridian, generally without being at all connected with the preceding portion of the work. A large compass fitted with plain sights, mounted on a ball and socket, or with parallel plates, and connected with a stand similar to that of a theodolite, forms this instrument. The internal part of the compass box is graduated very distinctly, sometimes into twice 180°, or four times 90°, as well as quite round the circle to 360°. At zero, or 360°, is fixed a perpendicular sight vane, with a fine wire strained on its opening: opposite to it, over the division 180°, is fixed a similar vane and strained wire; these wires being in the meridian or N. S. line of the compass box. A small orifice is also made in each sight vane, in a line with the strained wire, to which the eye is applied on taking an observation, the object being bisected by the perpendicular wire, thus:—at any station from whence it is intended to carry out a line, set up the instrument, and by means of the ball and socket and spirit-bubbles set it level; or, if the instrument has no spirit-bubbles attached to it, notice if the needle plays freely, by which means it may be set up very nearly level; clamp the ball and socket tight, and turn the sights which are fixed to the compass box in the direction of the proposed line, which, when bisected, will give the desired angle, by noting exactly what degree the north point of the needle becomes stationary opposite. As the correctness of this instrument entirely depends on the fine performance of the needle, it will be requisite to have a very sensitive one, and great care must be taken that the instrument is not stirred after bisecting the object, and previous to noting the angle. The needle should always be thrown off its centre, and not allowed to play except when in the act of using it, otherwise the fine point on which it revolves, and its accuracy depends, will soon be destroyed. The surveyor should also notice that in this instrument, and, generally, in all modern instruments, the compass is marked contrary to nature; that is, the West is substituted for the East, the North-East for the South-West, &c.: by thus altering the cardinal points the readings of the needle show the actual direction of the line. When this artifice is not resorted to, it will evidently be necessary to graduate the protractor, by which the survey is plotted, the *reverse* way of the instrument.

It may be useful to know that electricity is excited by rubbing the glass of a compass box with a silk handkerchief to remove wet or dirt, and in such state affects the needle; it is however at once discharged by touching the glass with the wetted finger. For a further description and method of using this instrument, and plotting the work, see "subterranean surveying" and "prismatic compass."

The Prismatic Compass is a very useful little instrument, and is employed in the same manner as the circumferenter, that is, all the angles are taken from the magnetic meridian; but, instead of a needle with the divisions on the compass box, it has a graduated floating card attached to the needle, similar to a mariner's compass. This card is usually graduated to 15′ of a degree, but angles cannot be taken with it to less than about 30′, or half a degree, which renders it very unfit to be used, except in filling in the detail of a survey, where the principal points have been accurately determined by triangulation, or by means of a theodolite, where great accuracy is not required, or the scale of the required plan very minute. The graduations on the floating card commence at the North point, or zero, and are numbered 5°, 10°, &c., round the circle, to 360°. Attached to the instrument is a perpendicular sight vane, with a fine thread or hair strained along its opening, and opposite to which is the prism. On applying the eye to an orifice over the prism, and bisecting any object with the thread in the sight vane, the division on the card coinciding with the thread—which will be the angle formed by the object with the meridian,—will be reflected to the eye of the observer; but care should always be taken that the card is quite stationary at the time the angle is noted. Some attention will also be necessary in holding the instrument in such a position that the card will play freely on its centre, otherwise the results obtained will be liable to great inaccuracy.

The angle formed by one object with another may be easily ascertained with this instrument, by first finding the angle formed by each with the meridian; the difference will be the angle required: thus,—suppose the bearing of one object is found to be 20°, and the bearing of another 45°; the angle subtended by these two objects will be 25°, which is the less subtracted from the greater: but where the difference is greater than 180° it must be subtracted from 360°, as, in the following example,—the bearing of one object is found to be 345°, of another 30°; the angle subtended in this case will be the difference between 30° and 345°, = to 315° subtracted from 360°, = to 45°. But the best method of using this instrument is to keep each angle separate, noting them all in the field-book from the meridian; there will then be less liability to error, and the plotting be performed with great rapidity, by means of a protractor formed on the same sheet of paper as the intended plan, and in the following manner:—if only filling in the detail of a map, the magnetic meridian will have been previously determined, if not, draw one, taking care that it is so placed that the paper will take in the greatest extent of survey. A line is then to be drawn in a vacant corner of the map, parallel with this meridian, and from a metallic protractor mark off the degrees and half degrees, which may be done in a few minutes: number them as 10°, 20°, &c., to 360°, commencing from the North point as on the floating card. Then, to lay off an angle from the meridian, apply the edge of a parallel ruler to the centre of the circle forming the protractor and the required angle, and slide it parallel to itself to the point from whence it was taken, draw in the line and mark off the distance; proceed in this manner until all the required angles are laid down on the plan. If the survey is very extensive, describe two or three of these protractors on different parts of the plan, taking great care that the meridian lines are exactly parallel. A much more convenient method than the above of laying down angles taken from the meridian is, to make a protractor on a separate sheet of paper or card-board in the manner just described; then cut out the blank paper withinside the circle on which the degrees are marked, and it will prove a most convenient protractor, and may be attached to any part of the map, and the angles laid off as before by means of the parallel rule. Or the meridian line being drawn through any point on the plan from whence angles have been taken, such angles can be instantly marked off by laying this paper protractor on the meridian line, observing that the N. S. line on the protractor must exactly coincide with that on the plan. The same method is also observed in plotting surveys made with the circumferenter, when the compass box is marked in the manner we have pointed out in our description of that instrument.

The Box or Pocket Sextant is the most useful instrument a surveyor can have, and sufficiently accurate for all purposes of surveying, except, in setting out long lines, or laying out large triangles in a hilly country, in which case a good theodolite is indispensable; but, for filling in the detail of a survey, it is unrivalled both for accuracy and expedition. The pocket sextant is usually divided on silver to 140°, although a greater angle than about 110° should not be taken with it at once, but angles of 220° or less may be taken in twice, in the following manner. Set up a mark at as great an angle as can conveniently be taken at once—say 110°; then turn half round and take the remainder of the angle, which we will also call 110°; these two angles added together will make 220°. It is thus perceived that angles of any magnitude may be taken with the sextant; in three times the entire circle may be taken; but an angle beyond 110° should, if possible, be avoided, as with a little attention and trouble the *supplement* of an angle when greater than that may generally be obtained.* A small telescope is sometimes attached to assist the sight, but is very inconvenient, taking up more time to arrange for distinct vision than for taking the angle, but there is a small aperture opposite to the half silvered or horizon glass to take plain sights with, in place of the telescope, which is far preferable. It should be particularly observed, that the objects or marks to which angles are taken should be as far from the observer as possible, as, if very near, little dependence can be placed on the observations; this direction should be carefully attended to when adjusting the instrument. The method of taking angles with the sextant is for the observer to place himself on a line at the exact point from which it is intended carrying out another line, the angle of which with the preceding, he is desirous of knowing. Take the sextant in the right hand, the case of the instrument forming a handle, and apply the eye to the small aperture, and look through the unsilvered part of the horizon glass to some object or station mark on the first line; then, with the left hand, keep turning the milled head screw, which carries the silvered glass, until it reflects an object on the second line, in the silvered part of the horizon glass; and when the two

* Angles of 180° are necessary to prove its correctness of graduation, although an angle of 0° degrees will prove the accuracy of position of the glasses.

objects are in exact conjunction (that is, the object viewed direct through the unsilvered part of the horizon glass, and the reflected object in the silvered glass, appearing as one), the desired angle will be given. From the construction of this instrument angles can only be taken accurately when the objects and observer are on the same horizontal plane, for the angle subtended by two objects becomes less and less the further we recede from them, and when the objects observed are not on the same plane with the observer, the difference of distance will be as the hypothenuse to the base of a right-angled triangle, and, consequently, in every such case the observed angle will be less than the real angle. A very good method for correcting this error, when the sextant is used in a hilly district, is, to suspend plumb lines in the direction of the objects to be observed from the angular points, by two or three crossed sticks, and for the observer to sit on the ground, and hold the instrument exactly over such points, observing the angles subtended by the several plumb lines, instead of the objects direct.*

The method of adjusting this instrument is exceedingly easy and correct: first, observe some well defined vertical object through the unsilvered part of the horizon glass, and turn the milled head which moves the silvered glass, until the same object is reflected; *i. e.*, the observed and reflected object,—which in this case will be the same,—appearing as one; the index or vernier will then stand at 0° on the limb, if the instrument is correct, if not, the reading on the limb will show the error of the instrument. Or the vernier being set at 0°, the object viewed direct and by reflection will appear as one if the instrument is in adjustment, if not, it will overlap and appear unconnected. To remedy this, the

* To determine heights with the sextant, see an advanced part of our work.

horizon glass must be altered — by means of a screw on the circumference of the instrument,—until the object viewed direct and by reflection appear as it really is—one, the vernier standing at 0° on the limb. Next, look at the distant horizon, and if viewed direct and by reflection, it is in coincidence, the instrument will be, in perfect adjustment, otherwise, move the screw on the face until it is perfected. A key to fit both screws is fitted into a spare place in the face; the angle is read off with the assistance of an attached lens, with a hinge joint, moveable over every part of the limb and vernier.

The Theodolite is the best of all angular instruments for surveying; the improvements it has received from time to time by various surveyors rendering it almost perfect. The theodolites in general use have telescopes mounted in Y *s*, similar to that of the Y level, with the attached spirit-bubble. Near the eye end of the telescope is a flat ring — moveable by means of four screws on the outside—on which are fixed cross hairs, generally formed of spider's web; the intersection of these cross hairs is the optical axis of the instrument, technically termed the line of collimation, and the arrangement, a diaphragm. A semicircular arc for taking vertical angles is fixed to the Y *s*, and from the centre of the circle describing this arc, are projecting arms resting on standards, — which are fixed to the upper horizontal plate on which the verniers are marked, — the whole moving in a vertical plane. The chamfered edge of the lower plate is divided quite round the circle into 360°, and numbered 10°, 20°, &c., to 360°—the intermediates, as 5°, 15°, 25°, &c., being represented by longer divisions than the single degrees, and the half degrees, into which it is generally divided, by shorter lines than the degrees; the plates together are called the limb. For the purpose of bisecting objects correctly, a slow motion screw is attached to the upper plate, the clamp screw securing it when the object is nearly bisected, the slow motion screw then moving the vernier through the least possible space, perfects the bisection. In good instruments, similar clamp and slow motion screws are attached to the lower plate, by which means (as will be presently explained) angles may be repeated any number of times, thereby insuring the degree of accuracy required. A vernier is fixed to the upper plate, through which the semicircular arc for taking vertical angles passes; two spirit bubbles are also attached to this plate at right angles to each other, for the purpose of setting the instrument horizontal. A small compass is also fixed on the upper plate, the N. S. line of the compass box ranging with the line of sight: its use is in noting the bearings at the different stations as a check on the angles.* The whole is mounted on parallel plates similar to those of a level.

The principal adjustments requisite to this instrument are the same as those for the Y level (to which the reader is particularly referred); *viz.*, that the line of collimation, or optical axis, of the telescope must coincide with the cylindrical rings on which it turns in the Y *s*, and that the bubble must be parallel to this optical axis. The method of performing these adjustments is very fully explained in our account of the Y level. The next adjustment is to make the axis of the horizontal plates truly vertical: to do which, set up the instrument as near level as may be, the telescope lying over two of the parallel plate-screws, and

* In setting out a long base line, the extremities of which are beyond vision, the bearing should be taken at the commencement, at the termination, and at several intermediate points, as a check on its straightness.

by means of the clamp and slow-motion screws attached to the vertical arc, bring the bubble connected with the telescope into the centre of the tube. Then move the instrument half round, and the telescope will be over the other pair of plate-screws: if the bubble then remains in the centre of the tube, it will be right; if not, correct it—one half by the clamp and slow-motion screw attached to the vertical arc, and the other half by the parallel plate-screws. If the bubble does not then remain in the centre of the tube, while the instrument is turned quite round, it must be repeated until the result is satisfactory; when, if the bubbles on the vernier-plate are correct, they will also stand in the centre of their tubes: if not, the screws at either end—connecting the bubble-tubes with the vernier-plate—must be altered as much as will bring them to the centre of their runs. The adjustment necessary to the vertical arc may now be attended to: if the vernier stands at zero, when the former adjustments are perfect, it will be correct; if not, alter the vernier by means of the screws attaching it to the plate until it is in such position, or otherwise note the error, and allow for it in each vertical angle. But, for the purpose of accurately noting this error, it will be necessary to take the vertical angle of some conspicuous object with the telescope, reversed in the Y *s*; to do which, the limb must be turned half round, when the telescope will occupy the same position as at first, but the vertical arc will be reversed: the mean of these two readings will be the amount of error, which we should advise the surveyor to correct, in preference to allowing for it at each vertical angle, as being less liable to error.

Instrumental parallax is often the cause of as much

perplexity in surveying instruments as in those for levelling; for the method of procedure to correct this, the reader is referred to the account of levelling instruments in another part of our treatise; where, also, he will see the method of replacing the cross hairs of the diaphragm, in case he should happen to break them. The surveyor is strongly recommended to have a good solid stand for his theodolite, similar to that described for levelling instruments; he will thereby avoid the tremulous motion and attendant uncertainty consequent on the use of the round weak legs generally applied to surveying instruments, which appear to be made rather for show and convenience of carriage than use: a solid, firm, and immoveable mounting for the instrument being a desideratum when taking observations.

The method of observing angles with the theodolite* is, to set it up exactly over some station—which is easily done with the assistance of a line and plummet suspended exactly under its centre from a hook attached to the stand—and level it by means of the parallel plate-screws; † then clamp the upper and lower plates together, and turn the instrument towards some station mark. Clamp the lower plate when the object is brought nearly into a line with the optical axis of the

* When the angles of very distant stations are required, the best time for observing, in dry weather, is about an hour before sunset; in wet weather, early in the morning, if not misty. Important angles should not be taken, if it can be avoided, in the middle or towards the close of a hot day in damp weather. See "Refraction," in the introductory chapter to "Levelling."

† When levelling the instrument, the opposite parallel plate screws should be moved simultaneously, *i. e.*, one unscrewed and the other screwed at the same time, but one should never be tightened until its opposite is slacked. Only a very slight range of obliquity should be given to the parallel plates, and the screws worked but moderately tight, otherwise the thread will soon be destroyed, and consequently, the perfect action of the instrument. On the other hand, *slack* screws are as prejudical to the observations as tight screws to the mechanism of the instrument; a medium shows the master-hand.

telescope, and perfect the bisection with the slow-motion screw, observing in every case to bisect the object as near the ground as possible. Then read off the degrees and minutes, at which the upper plate stands on the lower, also the seconds if necessary, taking the mean of two or three verniers, if there are so many, but two should always be used. The *upper* plate may then be released (the lower one remaining clamped, care being taken that it receives not the least motion), and the telescope turned towards another station; then clamp the upper plate, and bisect the object with its own slow-motion screw; read the degrees, minutes, and seconds, as before, and their difference will be the desired angle. But, for general purposes, it will be found much easier, and be *less liable to error*, to set the vernier at 360° and clamp it to the lower plate; turn the instrument bodily round in the direction of the first object, clamp the lower plate, and bisect with its slow-motion screw; then release the upper or vernier-plate (the lower plate remaining clamped), and turn the instrument in the direction of the second station, clamp and bisect as before, and the angle read off will be that required. This method of taking angles will, in practice, generally be found the most preferable, although not so correct as the former, it being almost impossible to set the instrument exactly to 360°; but in extensive operations, and where many angles are taken from one station, the former method is greatly superior. The bearing—which is the magnetic angle, as pointed out by the compass-needle—may also be noted at each *principal* station, which will be a check on the accuracy of the angles; but, in the usual description of theodolites, the bearing cannot be read with any degree of accuracy, except by setting the *plates* and *needle* at zero, and reading the angle on the limb when the object is bisected, which will give the required bearing. This method of obtaining bearings is very much superior to that of reading direct from the compass-box, and should always be practised. Great care will, however, be necessary in levelling the instrument in each direction, otherwise the needle will play obliquely, and not give the correct bearing either in the compass-box or on the limb.

Vertical angles are taken for the purpose of reducing lines measured over steep ground to the horizontal measure. On one side of the vertical arc are engraved the requisite divisions for determining the angle, and on the other, the number of links to be deducted from each chain's length, to reduce it to the horizontal measurement. For the purpose of determining this angle, it is necessary to set up a mark at the exact height of the optical axis of the telescope, at the extreme point of the sloping ground to be measured, and on which the reduction is to take place; the requisite allowance can then be made in the field in the following manner:—suppose the angle of elevation is 14° 30′; on the other side of the arc will be found the figure 3, and a fractional part, which signifies that the chain requires to be lengthened 3 links and a fraction, or that the chain is to be drawn forward on the ground that quantity, to bring it to the horizontal measurement. In the common operations of surveying it will be found most advantageous to make the necessary allowance in the field, especially where many fences are crossed, or numerous offsets taken, as it will be found very troublesome in plotting to make the necessary reduction for each distance; but where great accuracy is required, the angle should be noted, and the necessary reduction made when plotting the work. Thus, on reference to the table for reducing hypothenusal lines to horizontal, it will be found that a reduction of 3.18 links must be made from each chain's length; now in the field, we can very well allow for the three links on each chain, but not for the fractional parts, which, on a line of any great extent, would make some difference. It should therefore depend on the description of work, whether the allowance be made in the field or in the office, the latter, as we have observed, being the most correct. In taking horizontal angles with the theodolite, the object is bisected with the vertical wire, but in taking vertical angles, the bisection is made with the horizontal wire; the latter is always the case in using levelling instruments, the former serving to show the operator when the staff is held perpendicular. In taking horizontal angles with the theodolite, we have found it a good plan, especially where several angles are taken from the same station, to read the angles back again, *i. e.*, not moving the limb but the vernier backwards, bisecting each object a second time, and observing if the angle reads the same as before. Further information respecting vertical angles will be found in our article on levelling with the theodolite.

Captain Everest's improved Theodolite differs considerably, and has many decided advantages over those in common use. It has been extensively used in India, and is now becoming very common in this country; this description of theodolite is generally bronzed. It is not thought necessary to give a minute description of this instrument, as five minutes' examination will more fully explain its advantages than the most lengthened account. This theodolite has three verniers for the horizontal angles, the mean of which is taken, as also of the two verniers for the vertical angles; and the whole instrument, in place of being mounted on parallel plates, is fixed on a tripod stand, three foot-screws serving to set it level. The method of adjusting this instrument, it will be perceived, is somewhat different from other theodolites.

To adjust the level, so as to make the axis of the horizontal limb truly vertical, bring the spirit-bubble attached to the bar over two of the foot-screws, and by their motion bring the bubble to the centre; then turn the instrument *half* round, when, if the bubble remains in the centre of the tube, it will be right; if not, correct half the error by raising or lowering the tube itself, and the other half by one of the foot-screws. For the line of collimation, bisect an object with the vertical wire, reverse the telescope, and the bisection will remain perfect if correct; if not, correct half the difference by the collimating screws, and the other half by a horizontal motion of the instrument. To correct the vertical arcs, take the altitude or depression of any object with the telescope reversed; half the sum will be the true angle to which the verniers must be set; and bring the level connecting the verniers to the centre of its run by the adjustment screws at either end.

We have a 5-inch theodolite (which is the most useful size for general purposes) of the above description, made by Troughton and Simms, which performs admirably, but have found it necessary to adopt parallel plates instead of the tripod (which had a very sensible lateral motion), retaining however the tripod, by which means we have the advantages of the instrument (such as being able to use it without a stand) without its defects. Over the verniers are engraved the letters A, B, and C; degrees and minutes being always read from A, and minutes only from B and C, and a mean taken as follows :—

On clamping the plates and bisecting the first object, the reading at

A was	7°	34′	-″
B ,,	-	33	-
C ,,	-	33	-
	7	33	20 mean.

On bisecting the second object whose angular distance from the first was required, the reading at

A was	45°	59′	-″
B ,,	-	58	-
C ,,	-	59	-
	45	58	40
Subtract	7	33	20
Angle required	38	25	20

The vernier A may be set to 360° in the same manner as described for the common theodolite, and the reading, when an object is bisected, will be the desired angle. Neither is it necessary to read the three verniers, if the description of work does not require such accuracy. Theodolites of the above size and description will be found the most useful for all ordinary purposes of surveying; and, moreover, are so light as to be carried about in the field without inconvenience.

Some few theodolites have been made for professional gentlemen, on the principle of a Transit Instrument, *i. e.*, instead of having a semicircular arc for observing vertical angles, an entire circle is applied, and the telescope—the axis of which is more above the horizontal plates than in the usual instrument—moves in a vertical plane, quite round the circle. Of course, an instrument so constructed presents the greatest facility in ranging out straight lines. We understand that the whole of the difficult and complex works connected with the South-Eastern Railway, at Dover, under the able direction of Mr. Wright, have been set out with such an instrument.

A further advantage attending the use of a theodolite in surveying, beyond determining the position of lines by angular measurement, is, in the being able to convert it into a Transit Instrument, and, thereby accurately laying out right lines in positions where the unaided eye — or merely assisted with a plumb line — would be quite unequal to the task. The method of using a theodolite under such circumstances is, to set it up over the angular point, carefully level the plates, and determine the direction of the required line; when this is done, the assistant is to pass along the intended line and set up a few ranging poles, or common whites, being motioned by the principal until such marks are exactly in the line of sight. When these marks have been set up for as great a distance as the instrument will command, or the assistant can perceive the signals, remove the theodolite to a spot beyond the *farthest* mark — leaving a ranging pole with a flag or other conspicuous mark at the angular point. Place the theodolite at the new station, as near on the line as possible, turn the telescope in the direction of the line backwards, level the plates, — *transversely* only,—and observe if all the marks previously set up are bisected by the vertical wire; if not, move the stand of the instrument gently to one side or the other, levelling the plates at each alteration, until all the back marks are covered by the vertical wire. When this is accomplished, open the clips confining the telescope — the plates remaining clamped — and reverse it, end for end, when fresh marks may be set up as at first, this process being repeated until a sufficient distance is ranged out. Another, and perhaps better method for accomplishing this last mentioned object, especially where common instruments are used, is, not to reverse the instrument, but to keep advancing with the sights; the theodolite at each change being set up behind at least three marks, fixed from the previous station; or, otherwise, after setting out a portion of the line from the angular point, the theodolite may be carried far beyond the extreme mark, and placed in the line in the manner previously described, and the back marks bisected. Fresh marks may then be planted quite up to the instrument, when, instead of reversing it, it is to be removed far onward as before, and the line brought on in the manner just described.

The various other Instruments which have been employed for taking angles, and for general surveying purposes, such as the plane table, semicircle, cross staff, triangle, and other similar contrivances, we do not think it necessary to mention, further than that they are all, at the present time, nearly, if not altogether exploded.

There are, however, a few other contrivances in use for facilitating surveying operations, which we think the present a proper place for introducing. First among these is the Optical Square, an instrument for setting off right angles. Its construction is very simple, being merely a metal case with the two glasses of the sextant fixed, and reflecting a constant angle of 90°; its size is such as to allow of being carried in the waistcoat pocket without inconvenience, and its use the greatest possible in setting out all kinds of work, taking long offsets, and other similar operations.

The Perambulator or measuring wheel is useful in measuring circuitous distances, such as crooked roads, &c., but corrections will have to be applied in case the ground undulates; a portable instrument of the same kind may often be used with advantage for determining circuitous distances on maps.

The Pedometer, although rarely applied to surveying purposes, presents great advantages in sketching ground, or in filling in the detail of a plan, especially in coast surveys. Its principle of action is simply that of registering the number of paces by wheel-work, which is done by fastening a lever in connection with the wheel-work to the thigh. In order to ascertain any distance, it is necessary to walk at a regular pace direct to the object, and the mean length of a single pace being previously ascertained, which, multiplied by the number of paces registered by the instrument, will give the total distance.

The Clinometer, or Batter-level, is another instrument of great use in surveying, although it is rarely or ever applied to such purposes. It is chiefly used by engineers for measuring the angles of slopes in excavations and embankments. But there is no reason whatever that it should not be applied by the surveyor to the determining the inclination of ground over which he measures his lines, and, consequently, enable him to make the requisite reduction without the incumbrance of a theodolite; and, from his being able almost without trouble to take a great many angles, where the ground is at all variable, the mean rate of inclination may be obtained with equal —if not superior—accuracy than when a theodolite is employed. The instrument consists of a small quadrant with an attached bar, to which a rod is fixed when in use; the quadrant has an index carrying a spirit level, moveable round the centre of the instrument. When in use, the rod is laid on the slope, and the index moved by the hand until the bubble assumes a central position in the tube; the angle then denoted by the index is the inclination. The ratio of slopes for the various degrees of inclination will be found at the end of our treatise, although in this part of our work it will scarcely be required.

An improvement on this instrument has been effected by a friend of ours;* as improved it is termed an Angle Meter and is represented in the accompanying sketch.

It will be perceived that this instrument has two

* Mr. J. R. Bakewell, C.E. It is manufactured by Cox, Barbican.

arms, and that the spirit level can be applied externally to one arm, and internally to the other, therein differing from the common Clinometer. But this difference is most important, as it enables us to take angles from above or below with equal facility, by merely transferring the spirit level, and, as this instrument is constructed with the greatest accuracy, and the quadrant engine divided, the results will be found sufficiently near the truth for all practical purposes. It may be used with advantage in practical engineering, for determining angles of strata, of constructions, and very many other purposes. To take the dip or under strata, or the angle of the under part of a roof, the spirit level remains on the lower arm of the instrument; but, for taking the dip or angle from the top, the spirit level is fixed upon the upper arm, the angle in either case when the other arm is brought to the required degree of inclination, being read off from the quadrant at the joint. A table of the ratio of inclinations for various angles is engraved on one arm, thereby rendering the whole complete. The small compass attached to the instrument adapts it to the miner's use, in taking the bearing, as well as the dip of mineral lodes, &c.

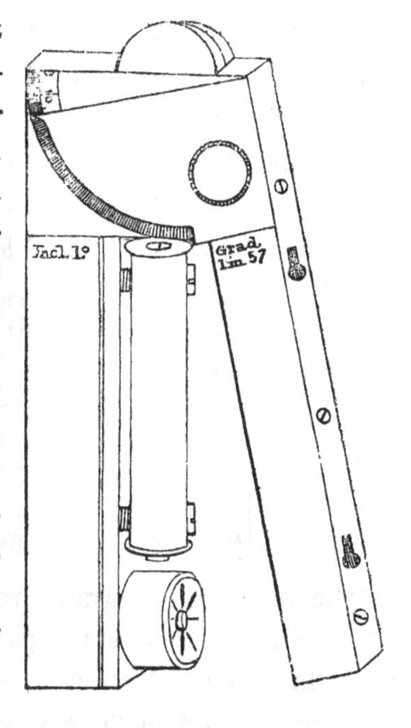

As a substitute for flags* or pieces of white paper

* White flags can be seen at a much greater distance than when of any other colour; red flags at a very short distance appear black, and are scarcely distinguishable with any background.

attached to the rods or poles used to define stations, we may mention the following contrivance.*

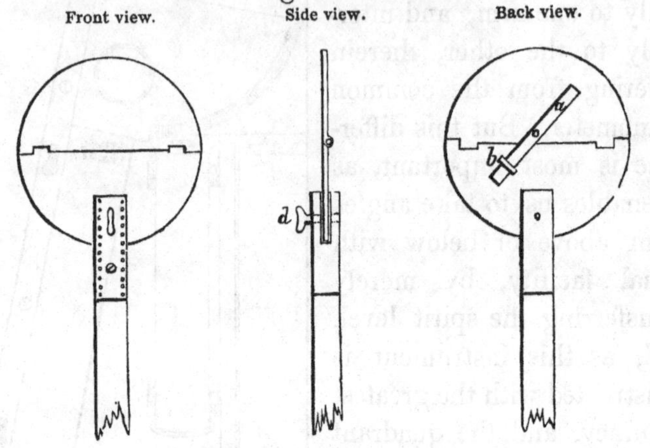

The preceding diagrams represent a disc of tin about six inches in diameter, painted white on both sides, and which, for convenience of carriage, is joined across the middle; its open position being secured by a little bar, *a*, to be turned into the catch, *b*. An iron ring or socket is screwed on to the top of the pole, and receives the disc in a slit, while the screw, *d*, secures it.

The following contrivance of a friend of ours,† for the same purpose is worthy of attention: it is represented in the following diagram: *a b* are two vanes of tin about six inches in length, and joined at the narrow part, *c*, where a socket receives a pin fixed on the head of the rod; the two vanes are bent or curved reverse ways at their extremities, which, being the broadest surfaces, hold the wind, and would be in equilibrio if the surfaces were plane, but ready to turn on the slightest preponderance being given to either side. This preponderance is effected by the concave extremity holding the wind, while it glides

* Communicated to the Civil Engineers' Journal by Mr. Dempsey.

† Mr. Robert Richardson.

off on the convex side; by which means a rotatory motion is produced, its rapidity according with the curve of the vanes. It is painted altogether white or half red and half white horizontally, and forms a very conspicuous object at considerable distances.

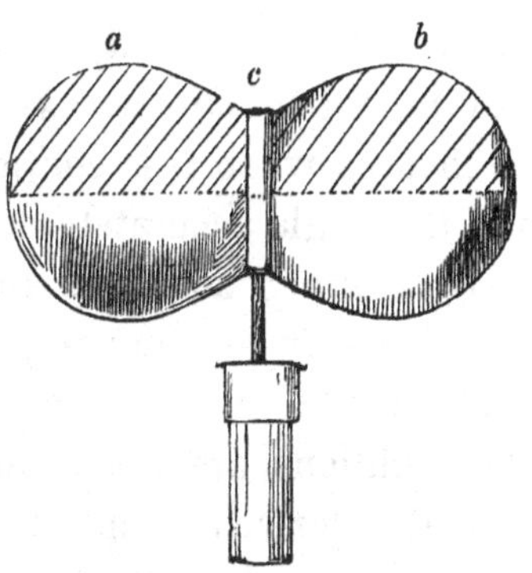

At the conclusion of this our treatise on surveying, we have to add a few remarks and observations which have been omitted in their proper places.

In surveys of the most preliminary kind we would always advise longitudinal side lines to be measured as a check on the base, and not, as is frequently the case, merely take the bearing or angle of the cross fences off the base line, and draw them out. We have often known mistakes in the length of a plan to be made from this cause, and in a recent survey, in which we were concerned, an assistant had filled in the detail in the manner just mentioned. By so doing, an error of ten chains occurred in measuring across a large enclosure, and which, from the absence of side lines, remained undetected until the distances were again measured, on taking a section of the country. In every instance where distances are not checked by side lines, the measurements should be made with especial care.

The direction of lines must of course always be determined by the nature of the ground and position of the objects to be delineated, but, by following the accompanying direction, great facility will be experienced in completing measurements, determining angles, &c.: it is to choose fixed marks to run the lines to, *beyond* the terminating point, by which means the measurement can be taken *on* the exact line, (which cannot always be insured when it is continued past the observed object), and the angle accurately determined without difficulty.

In making additions to an old survey it will always be necessary to compare, with the most scrupulous exactness, the scale of the old plan with that with which the additions are about to be plotted; and if in error, a scale derived from the map must be used in place of the original scale, to which it may have been plotted. And where the additions are required to be made with very extreme accuracy, we would recommend the length on a well defined portion of the map to be compared with actual measurement, and a scale to be directly deduced therefrom. Where any difficulty is experienced in the selection of well defined portions of the plan to commence operations from on the ground, we would recommend the end of a fence abutting on another nearly at right angles to be taken as a primary mark, and which point may, in all cases where the original survey is accurate, be determined within a link of the truth.

In computing the area of land we should have remarked in our previous observations that the horizontal superficial measurement is invariably given, and not the actual surface measurement; but, although we implied this meaning when speaking of the reduction of lines measured over undulating ground to the horizontal measurement, we think it requires further remark, especially as some extraordinary misconceptions are abroad respecting this matter. In a recent notice issued, requiring tenders for surveying the Isle of Wight, it stated that the survey should be trigonometrical, and that the horizontal area and *surface* measurement of each enclosure would be insisted on! Whether they have found a surveyor to accomplish the undertaking we are not aware, but certain we are of this, that no persons but the clever concocters of the circular in question would be able to instruct in the matter, in case it should be required. It is an absurdity to think of oblique superficial measurement, as the earth's surface is never disposed in a regular curve, and without it could be reduced to such form, all attempts would end in gross error. Besides, no more vegetable matter will grow on the sides of a hill than would grow on the area occupied by its base, and, therefore, in addition to its being an impossibility to correctly compute the surface area, it would be an injustice on the occupier of the soil if attempted. In surveying mineral property, where cubic contents are required, the surface measurement is then to be taken notice of, but it cannot be laid down on the plan or made use of in any way without very numerous vertical sections of the ground are taken, when an approximate cubit content may be found, and also a distant approach to the superficial area; but this part of the business is scarcely within the surveyor's province, certainly not when he is employed to measure the *area* of any district.

When a detail survey for a railway is required, it will be found a good plan, first to set out the centre line, and then, in filling in the detail of the plan, to take up the curves and every other part of the centre line in the same manner as a fence, but with more care. It will also be found of great service in such plans, if the surveyor is directed to sketch in the inclinations of the ground, and where the ground is varied, or of considerable inclination, for sections to be taken by the engineer, and plotted on the plan in their proper positions; for all engineering purposes, such a document would be altogether superior to a model.

When a survey for a railway is completed, the area of the land required from each enclosure, and also the area of the severance, should be written distinctly thereon, with the description of land, as arable, pasture, &c., also, the owner and occupier's names, which will be found infinitely superior to the usual detached reference books, or "terriers," as they are termed in some parts of England. In the reduced parliamentary plans for depositing, of course, a detached reference book is indispensable; it will also be found of great assistance to the surveyor or engineer in the preliminary plans to have each separate property coloured differently.

Where lines are measured over a gate or wooden fence, note the chainage, although it may not be a boundary, and cut a notch on the exact line, which mark can then be used as a false station, or it may become useful in retracing the line, if required, at any future time.

In our remarks on measuring inaccessible distances, we should have stated that it is a good plan to make the inaccessible distance of some integral length, as five or ten chains, or more if required, and to discontinue the direct measurement also at some integral length, by which means no confusion or error will be likely to arise in adding the distances together in continuing the measurement forward. Thus, if in measuring a line, an obstruction exists at 84 chains, stop at 80 chains, and let the unmeasured or inaccessible distance be exactly 10 chains, then it is evident that no mistake can possibly arise in the chainage, and it will be an easy matter to deduct the *back* measurement to the obstruction from the 90 chains, and add the *forward* measurement to the 80 chains, to correctly fix its position.

The last observation which we have to make is one of the greatest importance, and from the want of attention to which, immense sums of money have been fruitlessly wasted in the carrying into execution very many of our public works, especially railways. It is the connection of separate surveys, which has been hitherto so bunglingly done, that some degree of inaccuracy of position is at least generally expected at such points. But, how could it be otherwise, when so little attention has been given to the subject? Take, for example, 30 miles of country to be mapped by three surveyors, the usual method in such cases is, to give each, say ten miles, bounded by a road or river at the extremities, which distance they immediately proceed to complete, without reference to each other's operations. Now, when the three separate portions are completed, how are they to be placed so that each occupies its true relative position? In such cases there will be nothing but the deflections of a road or brook on one plan fitted into corresponding deflections on the adjoining plan, to determine so important a point. Sometimes, indeed, surveyors attempt to connect their plans by extending the base line from one portion far into the adjoining portion, taking up in its measurement the various objects it intersects; such a method may, in some cases, answer pretty well, if the several points are very accurately determined on both plans. The method, however, which we constantly practise and recommend is, for the surveyors to meet at the point of junction of the adjoining portions of survey, and decide on a *mutual* base line for such portion of each survey; they may then commence and complete their operations without further communication. When such maps are completed the engineer only requires to bring the corresponding points together, and the portions of base on the two plans into one right line with a straight edge, when each will be certain to occupy its true relative position. In a case where two surveyors are working towards each other, if their base lines will not intersect by being produced for a short distance into one portion, a new base may be commenced a short distance from the boundary that will intersect, which, on being laid down on both plans, presents the same ready facility of accurate connection. In like manner when one person makes a continuous survey, but which, from its general change of direction, requires to be plotted on several separate sheets, the disjunction should not be at the termination of a base line, but some short distance from it, so that portions of the same base may occupy portions of separate sheets. Such plans may then be placed in their true relative positions in the same manner as just described.

The scales to which surveys are usually plotted are—

For small plots of land, towns, villages, &c., where improvements are projected .... ½ a chain to 1 chain to an inch.
For Railway plans and other extensive surveys ...... 2 chains to an inch.
A good scale for the above purpose, which has also the advantage of working to links or feet............ 2½ chains to an inch.
For extensive plans, as Parish surveys, &c., the best scale, where quantities have to be computed therefrom; and that adopted by the Tithe Commissioners .... 3 chains to an inch.
For Preliminary surveys, without much detail....4 or 5 chains to an inch.

Beyond this last scale, surveys are rarely or ever plotted, but when a plan is required on a smaller scale it is usually reduced from a larger size. This is generally the case with Parliamentary deposit plans for railways, &c., which are required to be not less than 20 chains to the inch, or four inches to the mile: to save expense, the minimum scale is invariably adopted.

CONCLUSION OF SURVEYING.

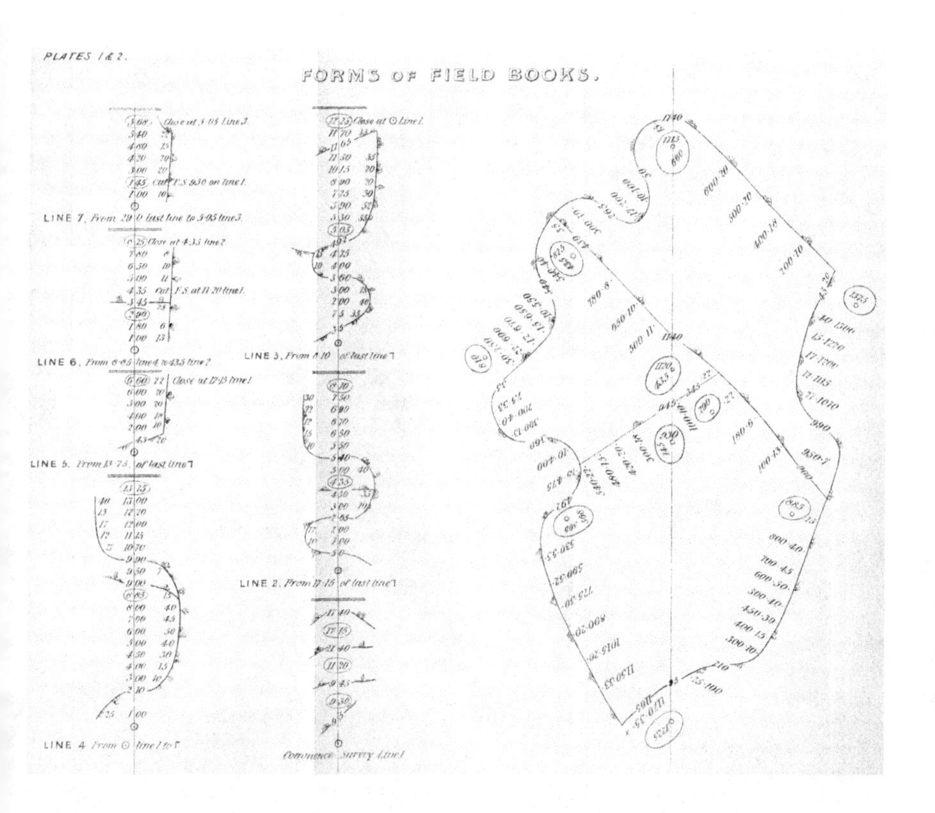

The material originally positioned here is too large for reproduction in this reissue. A PDF can be downloaded from the web address given on page iv of this book, by clicking on 'Resources Available'.

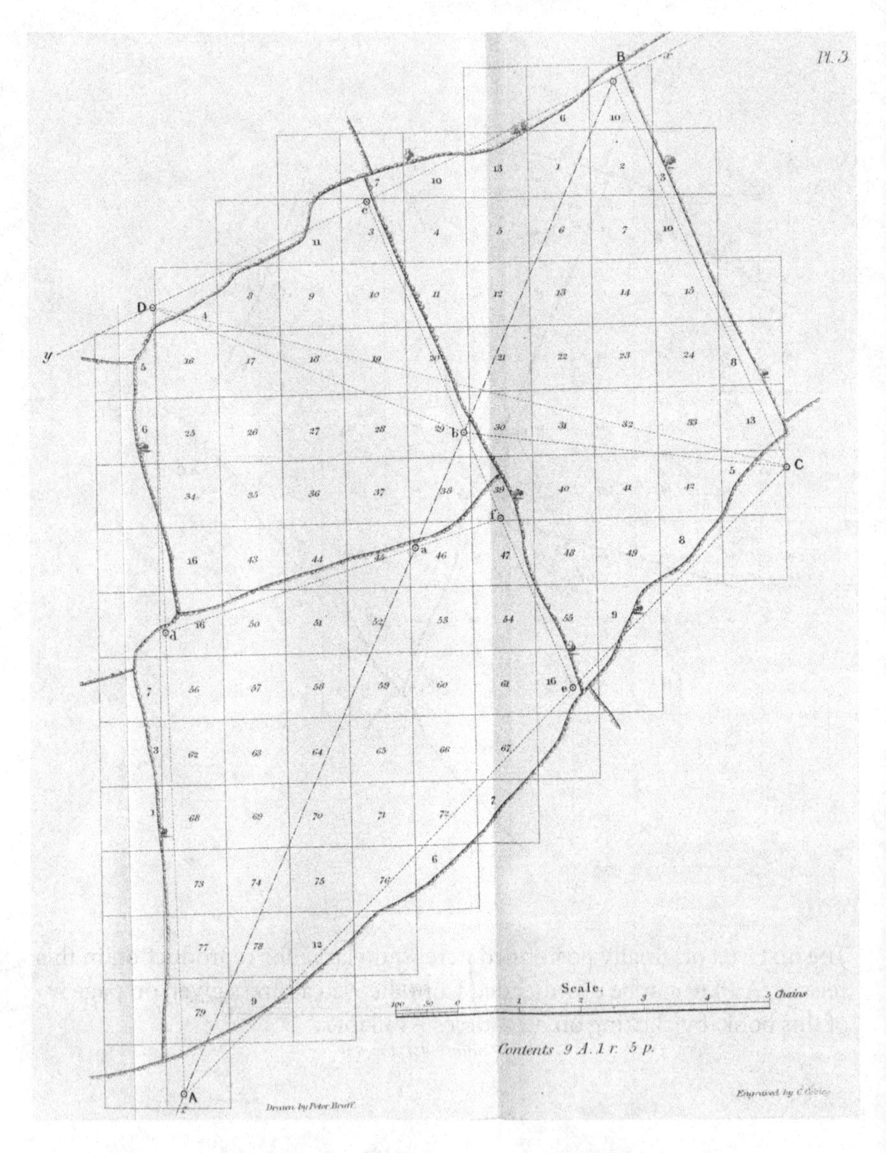

The material originally positioned here is too large for reproduction in this reissue. A PDF can be downloaded from the web address given on page iv of this book, by clicking on 'Resources Available'.

FIELD BOOK TO RAILWAY SURVEY, AT PLATE 7.

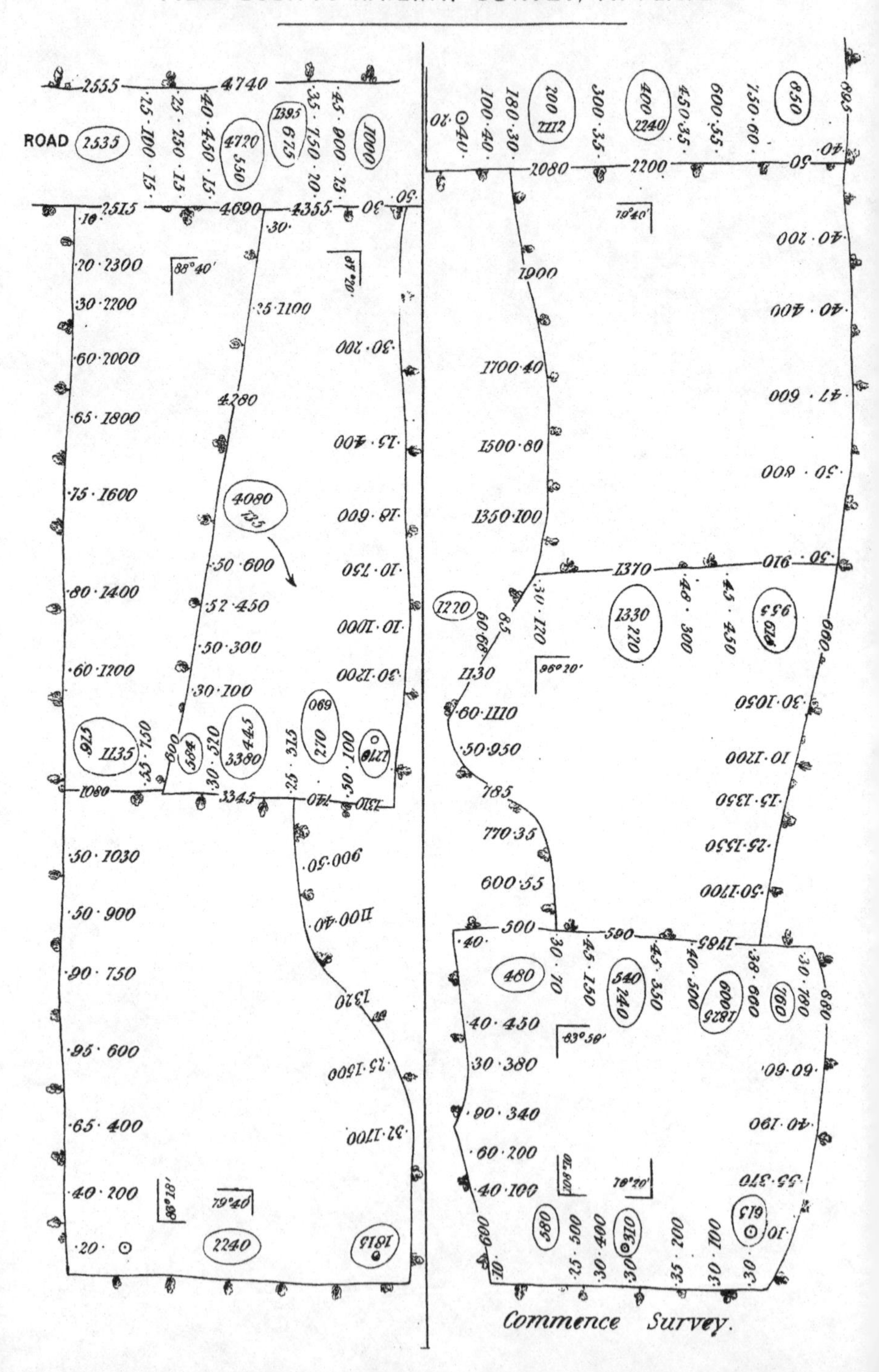

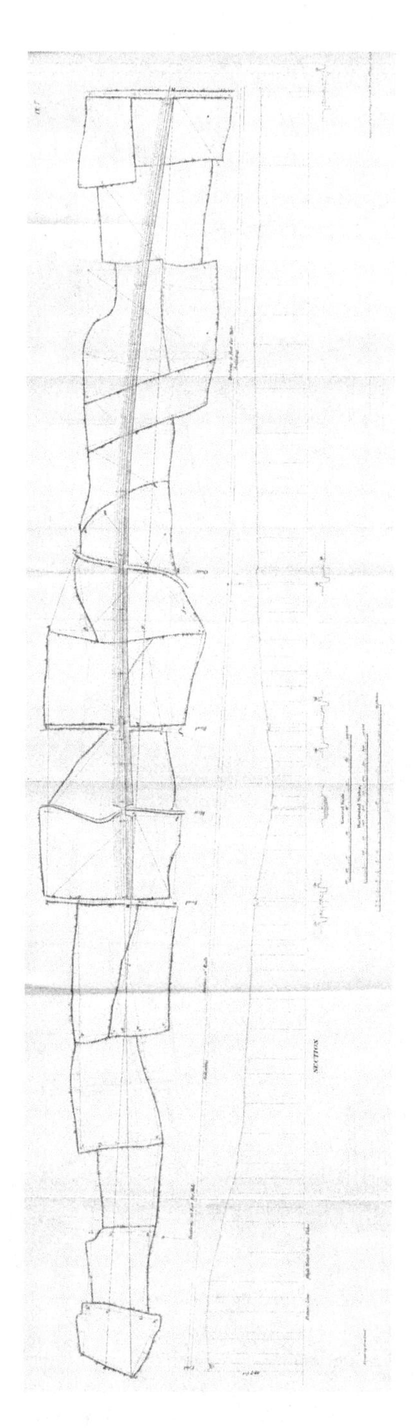

The material originally positioned here is too large for reproduction in this reissue. A PDF can be downloaded from the web address given on page iv of this book, by clicking on 'Resources Available'.

Printed in the United States
By Bookmasters